RISK ANALYSIS AND UNCERTAINTY IN
FLOOD DAMAGE
REDUCTION STUDIES

Committee on Risk-Based Analysis for Flood Damage Reduction

Water Science and Technology Board

Commission on Geosciences, Environment, and Resources

National Research Council

NATIONAL ACADEMY PRESS
Washington, D.C.

NOTICE: The project that is the subject of this report was approved by the Governing Board of the National Research Council, whose members are drawn from the councils of the National Academy of Sciences, the National Academy of Engineering, and the Institute of Medicine. The members of the committee responsible for the report were chosen for their special competencies and with regard for appropriate balance.

Support for this project was provided by the U.S. Army Corps of Engineers under DACW72-98-C-0003.

International Standard Book Number 0-309-07136-4

Library of Congress Catalog Card Number 00-108527

Risk Analysis and Uncertainty in Flood Damage Reduction Studies is available from the National Academy Press, 2101 Constitution Avenue, N.W., Washington, D.C. 20418; (800) 624-6242 or (202) 334-3313 (in the Washington metropolitan area); Internet <http://www.nap.edu>.

Sketch on the book cover is courtesy of the California State Library. It is a contemporary sketch of the city of Sacramento during the high water of the winter of 1849-1850.

THE NATIONAL ACADEMIES
Advisers to the Nation on Science, Engineering, and Medicine

National Academy of Sciences
National Academy of Engineering
Institute of Medicine
National Research Council

The **National Academy of Sciences** is a private, nonprofit, self-perpetuating society of distinguished scholars engaged in scientific and engineering research, dedicated to the furtherance of science and technology and to their use for the general welfare. Upon the authority of the charter granted to it by the Congress in 1863, the Academy has a mandate that requires it to advise the federal government on scientific and technical matters. Dr. Bruce M. Alberts is president of the National Academy of Sciences.

The **National Academy of Engineering** was established in 1964, under the charter of the National Academy of Sciences, as a parallel organization of outstanding engineers. It is autonomous in its administration and in the selection of its members, sharing with the National Academy of Sciences the responsibility for advising the federal government. The National Academy of Engineering also sponsors engineering programs aimed at meeting national needs, encourages education and research, and recognizes the superior achievement of engineers. Dr. William A. Wulf is president of the National Academy of Engineering.

The **Institute of Medicine** was established in 1970 by the National Academy of Sciences to secure the services of eminent members of appropriate professions in the examination of policy matters pertaining to the health of the public. The Institute acts under the responsibility given to the National Academy of Sciences by its congressional charter to be an adviser to the federal government and, upon its own initiative, to identify issues of medical care, research, and education. Dr. Kenneth I. Shine is president of the Institute of Medicine.

The **National Research Council** was organized by the National Academy of Sciences in 1916 to associate the broad community of science and technology with the Academy's purposes of furthering knowledge and advising the federal government. Functioning in accordance with general policies determined by the Academy, the Council has become the principal operating agency of both the National Academy of Sciences and the National Academy of Engineering in providing services to the government, the public, and the scientific and engineering communities. The Council is administered jointly by both Academies and the Institute of Medicine. Dr. Bruce M. Alberts and Dr. William A. Wulf are chairman and vice chairman, respectively, of the National Research Council.

Preface

Any review of the U.S. Army Corps of Engineers's approach to technical issues and their applications tends to be complicated because of the Corps's size, its lengthy and rich history, its relations with other federal agencies, and controversies that have followed the Corps for decades. This review and study were no different. Our study committee was challenged to analyze the Corps's risk analysis techniques in its flood damage reduction studies, a challenge that was magnified by the need to understand several related issues. Our committee experts in hydrology, engineering, and statistics found themselves analyzing not only risk analysis applications, but also considering levee certification policy and history, federal flood insurance programs, and U.S. floodplain management strategies. The committee undertook these peripheral investigations partly because of the need to adequately address its statement of task and partly out of intellectual curiosity. In any event, one implicit conclusion of our study is that an appreciation of the Corps's historical roles in addressing the nation's flood problems is necessary to understand the current issues the Corps faces in engineering and hydrologic applications.

The Corps's relatively new applications of risk analysis represent a significant departure from long-held, traditional approaches to addressing hydrologic, hydraulic, and geotechnical uncertainties. The former approach of adding freeboard to its levees was for several decades a sound strategy for coping with unquantifiable uncertainties. Because of historical momentum, this former approach has left a legacy that is not easily jettisoned. Several Corps of Engineers techniques and policies were based upon the concept of freeboard, and it will take some time for the agency to fully adjust to the new techniques.

In watching these changes within the agency, our committee gained an appreciation for the dedication of several Corps of Engineers staff members who assisted with this study. Much of the development of the risk analysis techniques has taken place at the Corps's Hydrologic Engineering Center in Davis, California. The committee expresses its gratitude and appreciation to Darryl Davis, the Center's director. Darryl has been a leader in promoting risk analysis applications within the Corps. The committee appreciates Darryl's frankness and cooperation during this study. David Goldman, also of the Hydrologic Engineering Center, has been central to tailoring the risk analysis techniques to Corps applications and deserves major credit for advancing risk analysis within the Corps.

Several other Corps of Engineers staff members shared their knowledge and views with the committee. Earl Eiker and Harry Kitch at Corps Headquarters in Washington, D.C., and David Moser of the Corps's Institute for Water Resources, spoke with the committee at its first meeting in Washington in December 1998. Staff from the Corps's Louisville district office hosted a visit by a committee member in the summer of 1999 and provided information for the committee's Beargrass Creek case study. The committee thanks Neil O'Leary, Richard Pruitt, and Matt Scheuler in the Louisville district office for their assistance.

The committee thanks Joe Countryman of MBK Consultants (Sacramento), Michael Grimm of the Federal Emergency Management Agency (Washington, D.C.), and Doug Plascencia of Kimley-Horn (Phoenix) and member of the Association of State Floodplain Managers—all of whom spoke with the committee at its second meeting in Davis in February 1999. Joe, Mike, and Doug provided compelling remarks that helped the committee consider wider implications of the Corps's use of risk analysis.

Peter Andrysiak, U.S. Army, and Mitchell Laird of the Louisville district also provided significant assistance in acquiring project documents and data.

The committee also thanks Stephen Parker, director of the Water Science and Technology Board (WSTB). Steve followed the progress of this committee closely, and the committee frequently drew upon his knowledge of risk analysis and the Corps of Engineers planning procedures. His experience in managing numerous WSTB reports was useful in helping the committee reach agreement on some key technical issues.

Finally, the entire committee expresses its gratitude to project assistant Ellen de Guzman. Ellen demonstrated superb organizational skills, reviewed and organized several drafts of the committee's report, and also showed a great deal of patience and aplomb in dealing with too many last-minute requests from the chair and study director.

This report has been reviewed by individuals chosen for their diverse perspectives and technical expertise, in accordance with procedures approved by the Report Review Committee of the National Research Council (NRC). The purpose of this independent review is to provide candid and critical comments that will assist the authors and the NRC in making the published report as sound as possible and to ensure that the report meets institutional standards for objectivity, evidence, and responsiveness to the study charge. The contents of the review and draft manuscripts remain confidential to protect the integrity of the deliberative process. We wish to thank the following individuals for their participation in the review of this report:

Paul Barton, U.S. Geological Survey
Leo Beard, Professor Emeritus, University of Texas
Stephen Burgess, University of Washington
John Cassidy, consultant, Concord, California
Susan Cutter, University of South Carolina
Des Hartford, British Columbia Hydro
Debra Knopman, Progressive Policy Institute
Eric Wood, Princeton University

Although the individuals listed above provided many constructive comments and suggestions, responsibility for the final content of this report rests solely with the authoring committee and the NRC.

GREGORY B. BAECHER
Chair

JEFFREY W. JACOBS
Study Director

Contents

Executive Summary

Reducing flood damages is a complex task that requires multidisciplinary understanding of the earth sciences and civil engineering. In addressing this task, the U.S. Army Corps of Engineers employs its expertise in hydrology, hydraulics, and geotechnical and structural engineering. Dams, levees, and other river works must be sized to local conditions; geotechnical theories and applications help ensure that structures will safely withstand potential hydraulic and seismic forces; and economic considerations must be balanced to ensure that reductions in flood damages are commensurate with project costs and associated impacts on social, economic, and environmental values.

Many flood damage reduction projects involve the construction of levees. The Corps's historical approach to coping with hydrologic and hydraulic uncertainties of large floods was based on a best estimate of the levee height required to withstand a given flood, which was then augmented by a standard increment of levee height called "freeboard." The best estimate has traditionally been based on the expected height of a design flood (e.g., a 100-year flood, the magnitude of which has a 1 percent chance of being equaled or exceeded in any given year, and which is here called the "1% flood"). Freeboard was then added above the expected height. Many Corps flood damage reduction projects used a standard of 3 feet of freeboard. "Three feet of freeboard" became an engineering tradition within the Corps and was employed in hundreds of Corps flood damage reduction studies and projects.

Challenges to the concept of a standard levee freeboard emerged in the early 1990s. For instance, it was noted that a standard freeboard did not account for geographic and hydrologic differences at different locations and may thus have provided different levels of flood protection in different localities. Procedures for calculating the economic benefits conferred by levee freeboard were also questioned.

The Corps felt that development and application of *risk analysis* techniques held great promise in addressing these issues, as these techniques aim to quantify and explicitly incorporate uncertainties in hydrologic, hydraulic, and geotechnical parameters into levee design analysis. It was envisioned that proper application of risk analysis could replace the need for a standard 3 feet of freeboard.

Risk analysis also became part of a federal levee certification procedure jointly conducted by the Corps and the Federal Emergency Management Agency (FEMA). Within the National Flood Insurance Program, FEMA identifies areas subject to varying degrees of flood risk on flood insurance rate maps. One of these areas is the Special Flood Hazard Area (SFHA), defined as the area that is inundated by a flood having a 1 percent chance of being equaled or exceeded in any given year (the 1% flood). Property within a Special Flood Hazard Area is subject to mandatory flood insurance purchase requirements and may be subject to local land use regulations, as well.

Floodplain property can avoid the Special Flood Hazard Area designation, and the mandatory flood insurance requirements that attend it, if it is protected by a levee certified to provide protection against the 1% flood. The Corps of Engineers is responsible for certifying levees as meeting this safety standard. As levee certification could exempt a community from flood insurance purchase requirements (and possible exemptions from local land use requirements), this certification procedure has great local economic and public policy significance.

The historical standard for levee certification had been that levees must provide protection to the average stage (height) of the 1% flood, plus 3 feet of freeboard. With the Corps's adoption of risk analysis techniques in the early 1990s, the freeboard standard for levee certification was abandoned in favor of the new risk analysis standard.

The public, however, was not entirely comfortable with the replacement of a time-tested standard by relatively new techniques. These issues came to a head in a Corps flood damage reduction project planning study in Portage, Wisconsin in the early 1990s. The Corps study recommended a levee of elevation 798.3 feet for the city of Portage. But this recommended levee (the "National Economic Development" levee project alternative) would not have been high enough to be certified as providing protection from the 1% flood. Because the calculations for levee height were based on the new risk analysis techniques, in 1993 the city of Portage, the Wisconsin Department of Natural Resources, and the Association of State Floodplain Managers challenged the study's results. An outcome of the ensuing discussions was that the U.S. Congress requested a National Academy of Sciences study of the Corps's use of risk analysis techniques. (The National Academy of Sciences was subsequently subsumed—in 1999—as part of the National Academies. The National Academies includes the National Academy of Sciences, the National Academy of Engineering, and the Institute of Medicine.)

The charge to the National Academies was included in the Water Resources Development Act of 1996 (WRDA 96) from the 104th Congress of the United States. Public Law 104-303 of WRDA 96 stated the following (Section 202h):

> The Secretary (Army) shall enter into an agreement with the National Academy of Sciences to conduct a study of the Corps of Engineers' use of risk-based analysis for the evaluation of hydrology, hydraulics, and economics in flood damage reduction studies. The study shall include—
>
> a) an evaluation of the impact of risk-based analysis on project formulation, project economic justification, and minimum engineering and safety standards; and
> b) a review of studies conducted using risk-based analysis to determine—
>> i) the scientific validity of applying risk-based analysis in these studies; and
>> ii) the impact of using risk-based analysis as it relates to current policy and procedures of the Corps of Engineers.

To carry out this assignment, the Water Science and Technology Board (WSTB) of the National Academies's National Research Council (NRC) appointed the Committee on Risk-Based Analyses for Flood Damage Reduction to conduct the study, with sponsorship provided by the U.S. Army Corps of Engineers.

APPLICATION OF RISK ANALYSIS TECHNIQUES

Recognition of engineering and economic uncertainties and their explicit quantification in flood damage reduction studies leads to projects that are better tailored to local conditions and available data than did the earlier, deterministic levee freeboard standard. **The new techniques are a significant step forward and the Corps should be greatly commended for embracing contemporary, but complicated, techniques and for departing from a traditional approach that has been overtaken by modern scientific advances.** While some technical issues are not yet fully resolved, the Corps has taken a significant step forward with the development of its risk analysis methods for flood damage reduction studies. The committee also notes that these advances have been made with a relatively modest investment of resources. There should be no turning back from this accomplishment.

The former approach of using standard levee freeboard did not provide consistent levels of flood protection across the nation. A consistent protection standard must properly account for local and regional differences in topography, hydrology, and hydraulics, which the standard freeboard approach did not. In some areas, for instance, as little as 2 feet of freeboard may be required to provide adequate flood protection, while in other areas, as much as 6 feet may be required. The traditional 3 feet of freeboard standard masks a significant degree of variation of risk of levee failure for citizens protected by these levees. This variation in risk of failure can be quantified by the Corps's new risk analysis procedure.

The committee divided its recommendations for improvement into the following areas: (1) refine methods relating to probabilistic and statistical modeling of floods, performance of flood damage reduction systems (e.g., levees), and flood damage assessment, (2) adopt a consistent terminology for communicating risk analysis concepts within the Corps and to the public, (3) simplify and improve the complex and somewhat confusing criteria for certifying levees for inclusion in the National Flood Insurance Program (NFIP), and (4) move toward a more comprehensive decision making approach in flood damage reduction studies.

Risk Measures and Modeling

The committee reviewed a computer program developed by the Corps's Hydrologic Engineering Center (HEC) in Davis, California. This computer program, the Hydrologic Engineering Center Flood Damage Assessment (HEC-FDA) program, is the principal tool used in Corps district offices to calculate flood damage risks. This program implements the Corps's risk analysis and builds upon a deterministic approach to flood damage estimation that evolved over several decades. Although it benefits from this experience, the Corps's risk analysis method suffers from the difficulty of translating deterministic practice into a risk analysis application. Assessing the risk analysis method's validity rests on the following questions:

1. Are the performance measures generated by the risk analysis method useful and complete?
2. Are all the important uncertainties included in the analysis?
3. Is the specification of these uncertainties proper?
4. Are probabilistic and statistical methods used correctly?

Risk analysis is applied to economic performance measures (project net benefits and benefit–cost ratio) and to engineering performance measures (probability of flooding). While the measures of economic performance in the new method are generally practical and informative, there are too many types of engineering performance measures to be clearly understood by most citizens. Standardization to one or two key measures of engineering performance measures would represent an improvement.

The committee recommends that the Corps use *annual exceedance probability* as the performance measure of engineering risk. This is a measure of the likelihood that people will be flooded (including the probability of failure of flood damage reduction structures, such as levees) in any given year, considering the full range of floods that can occur and all sources of uncertainty.

For engineering purposes, it is useful to calculate other system reliability measures, such as the conditional nonexceedance probability for the 1 percent (100-year) flood. But such measures are difficult to understand and are not as clear as the measure of annual probability of flooding, and they should not be used in communicating flood risks to the public.

Evaluation of the uncertainty in economic benefits that are attributed to knowledge uncertainties represents an important advance for the Corps. Such an evaluation is performed using a Monte Carlo procedure that evaluates expected annual damages using different possible parameter combinations for hydrologic, hydraulic, geotechnical, and economic models. While the Corps's conceptual approach to modeling flood hazard and associated damages—using relationships between flood frequency, stage–discharge, and damage–stage—is consistent with longstanding scientific understanding, certain improvements to this method are needed. Risk analysis measures for a project must rest upon complete and accurate specification of the uncertainties in each component of an analysis and upon correct probabilistic methods to quantify and combine those uncertainties. **As the current method has shortcomings in these areas, the committee recommends that the Corps improve its analysis of hydrologic, hydraulic, geotechnical, and economic uncertainties**.

The Corps's conceptual approach of distinguishing between natural variability and knowledge uncertainty is reflected in the engineering modeling components of its risk analysis method, but the conceptual approach needs refinement. Natural variability is variability assumed to be inherent in natural processes, such as flood frequencies or properties of geotechnical materials. Knowledge uncertainty is attributed to limitations of scientific understanding of natural processes. In some cases, such as the hydraulic relationship between a river's stage (height) and its discharge, the risk analysis method appears to include natural variability under knowledge uncertainties; in others, such as geotechnical levee performance, knowledge uncertainties appear to be included under natural variability. This is a critical issue, because knowledge uncertainties and natural variability each affect the calculations of risk in different ways.

The committee recommends that the Corps focus greater attention on the probabilistic issues of identifying, estimating, and combining uncertainties. Better specification of knowledge uncertainties in flood frequencies is needed. The uncertainty in the skewness coefficient for log-Pearson Type III distribution models of flood frequency (which are used by the Corps and other federal agencies for describing return periods of floods) should be explicitly included in the risk analysis. Some measure of the uncertainty inherent in computing flood–frequency curves from rainfall–runoff modeling is also needed.

The committee recommends that the Corps strive to reduce the considerable variation in the estimates of water surface elevation when using different models of river hydraulics. The Corps's experiences in applying alternative methods to estimate flood stage indicate that there can be substantial differences in the results.

The committee recommends that the Corps's risk analysis method evaluate the performance of a levee as a spatially distributed system. Geotechnical evaluation of a levee, which may be many miles long, should account for the potential of failure at any point along the levee during a flood. Such an analysis should consider multiple modes of levee failure (e.g., overtopping, embankment instability), correlation of embankment and foundation properties, hazards associated with flood stage (e.g., debris, waves, flood duration) and the potential for multiple levee section failures during a flood. The current procedure treats a levee within each damage reach as independent and distinct from one reach to the next. Further, within a reach, the analysis focuses on the portion of each levee that is most likely to fail. This does not provide a sufficient analysis of the performance of the entire levee. This has important implications for not only geotechnical and economic analysis of flood damages, but also for levee certification.

The Corps's new geotechnical reliability model would benefit greatly from field validation. The nation has many years of experience with levee performance and, unfortunately, also with levee failures. Much of this experience is documented and much is accessible to federal agencies. **The committee recommends that the Corps undertake statistical ex post studies to compare predictions of geotechnical levee failure probabilities made by the reliability model against frequencies of actual levee failures during floods. In addition, the committee recommends that the Corps conduct statistical ex post studies with respect to the performance of other flood damage reduction structures (e.g., embankments, detention basins, hydraulic facilities). These latter studies should be conducted in order to identify the vulnerabilities (failure modes) of these systems and to verify engineering reliability models.**

Economics

In the current Corps method (and as mandated by the federal *Principles and Guidelines*), flood damage is calculated for each set of project alternatives by aggregating over all existing structures (buildings) in the floodplain. Then, reduction in flood damage is calculated by taking the difference between this aggregate number and the corresponding aggregate damage without the project. Correlations among the random variables can introduce serious errors in the analysis. Each structure in a floodplain is modeled as if that structure exists in isolation from all others. The result is that the analysis incorrectly computes uncertainties associated with differences in economic damages that result from different project alternatives.

The committee recommends that the Corps calculate the risks associated with flooding, and the benefits of a flood damage reduc-

tion project, structure by structure, rather than conducting risk analysis on damage aggregated over groups of structures in damage reaches. Furthermore, the practice of summing and subtracting percentile values of probability distributions of flood damage in reaches to obtain risk measures of project economic performance is unsound and produces output measures of unknown accuracy. The outputs of the economic risk analysis using the current procedure are thus of questionable value.

CONSISTENT TERMINOLOGY

The committee noted that a variety of terms describing aspects of risk and uncertainty are often used interchangeably within and between the Corps's water resources programs. **The committee thus recommends that the Corps adopt a consistent vocabulary for describing risk analysis concepts, specifically distinguishing between risk, natural variability, knowledge uncertainty, and measures of system reliability**. The Corps should clearly distinguish between natural variability (based on the random nature of physical systems) and knowledge uncertainty (uncertainties attributable to limitations in the current state of knowledge).

LEVEE CERTIFICATION

In the early 1990s the Corps and FEMA adopted a risk analysis approach to replace the practice of certifying levees that had 3 feet of freeboard above the 1% flood level. This risk analysis approach and the levels of flood protection it provided were controversial. Negotiations between the Corps and FEMA led to the current practice of certifying a levee based on a three-tiered decision rule, using: (1) 3 feet of freeboard, (2) a conditional nonexceedance probability of 90 percent of passing a 1% flood, or (3) a conditional nonexceedance probability of 95 percent of passing a 1% flood. Although this three-tiered criterion represents a reasonable transition from the former certification criterion into the risk analysis framework, it has the following deficiencies: (1) it still leads to different levels of flood protection for different projects, (2) the three-tiered decision rule is unnecessarily complicated, (3) the method evaluates levees individually rather than as a levee system that is intended to provide flood protection for a community, and (4) certification is incomplete in that it considers only the 100-year flood, not the full range of floods.

The committee recommends that the federal levee certification program focus not upon some level of assurance of passing the 100-year flood, but rather upon "annual exceedance probability"—the probability that an area protected by a levee system will be flooded by any potential flood. This annual exceedance probability of flooding should include uncertainties derived from both natural variability and knowledge uncertainty.

The criterion for certifying a levee should be that it provides satisfactory protection against failure of the flood damage reduction system, expressed as an annual probability of flooding. This new criterion should promote better communication among the Corps, FEMA, other regulatory and expert groups, and communities and local cosponsors.

Substantial resources and time may be required to implement the annual exceedance probability approach for certifying a levee. Until the measure of annual exceedance probability is adopted as the key criterion for levee certification, the committee recommends that the Corps and FEMA set a single conditional nonexceedance probability for levee certification.

The former certification criterion was flawed in that it produced vastly different levels of flood protection for different communities. The committee recommends that the certification criterion provide a uniform level of flood protection. Which level of protection to choose is not obvious. Insisting on the highest level of protection would mean that only a small proportion of levees would be certified. In the committee's judgment, the certification criterion should be the level of protection provided to most people in the past—the median level historically provided. Based upon a small sample of all Corps flood damage reduction projects, the committee found that the median annual exceedance probability of Corps flood damage reduction projects is approximately 1/230.

This is the committee's best estimate of the median annual exceedance probability. To obtain a more reliable measure of the median annual exceedance probability of approved projects, the committee recommends that the Corps examine a larger number of flood damage reduction projects and audit the process of estimating the annual exceedance probability for these projects.

The committee recommends that the Corps develop a table showing percentiles of variability in the annual exceedance probability of its flood damage reduction projects. By choosing an appropriate percentile value in this range, a corresponding level of assurance can be obtained that the expected level of protection is at least 100 years, as required. It was the lack of allowance for this variability that led to the abandonment of the annual exceedance probability criterion during the 1990s.

FLOODPLAIN MANAGEMENT

Neither the U.S. Congress nor the Corps of Engineers have defined an explicit goal for management of the nation's floodplains. In the committee's opinion, the goal of floodplain management should be to use the land for the greatest social benefit. Broadening the scope of the Corps's risk analysis and expanding the types of alternatives considered would provide more useful insight about how best to achieve this goal.

As currently specified by the federal *Economic and Environmental Principles and Guidelines for Water and Related Land Resources Implementation Studies*, flood damage reduction studies emphasize direct economic damage reductions and the costs of alternatives; these are quantified in the Corps's risk analysis methodology. **To ensure that the Corps's flood damage reduction projects provide adequate social and environmental benefits, the committee recommends that the Corps explicitly address potential loss of life, other social consequences, and environmental consequences in its risk analysis.** Furthermore, the Corps's risk analysis should not be limited to structural alternatives such as levees, dikes, and dams. Nonstructural alternatives such as warning systems and zoning regulations should also be considered, both separately and in conjunction with structural alternatives.

Given the breadth of federal agencies and programs devoted to U.S. floodplain and flood hazard management, the Corps clearly cannot implement these recommendations alone. Further, it is not likely that such a broadening of the Corps's risk analysis methods will occur over a short period of time. To include a broader range of social and environmental implications in the benefit–cost calculations of flood damage reduction studies, appropriate revisions of existing legislation and planning guidance, consistent with these recommendations, may have to be enacted by the U.S. Congress.

To appropriately include such consequences and their relative importance, the committee recommends that the ecological, health, and other social effects of Corps flood damage reduction studies, and the tradeoffs between them, be quantified to the extent possible and

included in the National Economic Development Plan. More explicit efforts at including these types of consequences and values in the Corps's benefit–cost calculations should result in increased social benefits of the Corps's flood damage reduction studies. The Corps should seek guidance from the Office of Management and Budget and seek consistency with other federal agencies on the use of alternative metrics for incorporating potential loss of life, environmental impacts, and other effects of floods.

1

The Corps and U.S. Flood Damage Reduction Planning, Policies, and Programs

Organized in 1802, the U.S. Army Corps of Engineers has long been a key player in helping reduce flood damages in the United States. The Corps's role in addressing the nation's flood problems was solidified with passage of the Flood Control Act of 1936. This act established flood control (generally referred to today as "flood damage reduction" by the Corps) as a nationwide policy on navigable waters and their tributaries, and it deemed flood control as an appropriate activity for the federal government.

One of the Corps's primary means for helping reduce flood damages has been levee construction. The Corps has constructed thousands of miles of levees that reduce flood damages for hundreds of American cities and thousands of acres of farmland. Uncertainties in the frequency of floods, changes in land use, climate variability and change, and the structural and geotechnical performance of levee systems complicate the levee design process. Furthermore, levee certification criteria and federal flood insurance policies, factor into flood hazard mitigation strategies.

The Corps has long recognized that uncertainties affect levee design and performance. To account for uncertainties, the Corps has historically established levee heights to pass a flood of a given recurrence probability. This flood has often been the flood with a probability of 1/100 of being exceeded in any given year, commonly called the "100-year flood" and herein referred to as the 1%-flood. An increment of levee height, called "freeboard," was then added to the levee design height. This freeboard was intended to account for operational contingencies, level-of-protection assurance, embankment settlement, and the like. But in historical Corps guidance, freeboard was not considered as a "factor of safety," per se. As described by Huffman and Eiker (1991),

> "Conceptually, freeboard is provided to reasonably assure that the project design flow will be contained, given the uncertainty of water surface profile computation, and to minimize the damage and any threat to life in the event the levee is overtopped. In the design of freeboard it is convenient to consider that freeboard has two primary purposes (1) to achieve specific design objectives, and (2) to allow for the uncertainty inherent in the computation of a water surface profile."

A river at flood stage bears little resemblance to a lake on a calm day. It flows swiftly with rapids and waves, carrying trees, ice, and other flotsam. Waves or floating objects can overtop a levee, breeching it and causing flooding. Freeboard is a measure to prevent overtopping caused by higher water (river stage) than was forecast for the design flood, as some uncertainties may not have been explicitly considered. For decades, the Corps added 3 feet of freeboard to the design height of its levees, a principle that became a staple of Corps flood damage reduction studies and projects. The practice was also used by the Federal Emergency Management Agency (FEMA) in certifying levees under the National Flood Insurance Act of 1968 (which created the National Flood Insurance Program) and two subsequent revisions (the Flood Disaster Protection Act of 1973 and the National Flood Insurance Reform Act of 1994).

Many levees have been designed, built, and certified to a standard equal to the 100-year flood plus 3 feet of freeboard. Nonetheless, this approach has drawbacks, and Corps engineers and planners, flood damage reduction project cosponsors, and others called this traditional standard into question during the 1990s. In particular, this fixed-freeboard approach provided inconsistent degrees of flood protection to different communities and provided substantially different levels of protection in different regions. In a narrow channel with variable hydraulics, a 3 foot safety margin may yield an unacceptably high probability of being overtopped. In a broad basin with overflow into storage areas, 3 feet may result in an exceptionally low probability of being overtopped.

RISK ANALYSIS APPROACH

In the early 1990s the Corps began to explore an alternative analytical approach to fixed freeboard. This alternative, referred to as *risk-based analysis* (RBA) by the Corps, is more generally known as *risk analysis*. A risk analysis approach uses probabilistic descriptions of the uncertainty in estimates of important variables, including flood–frequency, stage–discharge, and stage–damage relationships, to compute probability distributions of potential flood damages. These computed estimates can be used to determine a levee height that provides a specified probability of containing a given flood.

The Corps unveiled its basic proposal for using risk analysis techniques to substitute for fixed levee freeboard at a 1991 workshop in Monticello, Minnesota (USACE, 1991a). In 1992, the Corps issued a draft engineering circular (EC) *Risk-Based Analysis for Evaluation of Hydrology/Hydraulics and Economics in Flood Damage Reduction Studies* (EC 1105-2-205). In 1994, the Corps updated this engineering circular. In March 1996, the Corps issued engineering regulation (ER) 1105-2-101, which represents current Corps policy and procedures for risk analysis (USACE, 1996a). The Corps currently employs risk analysis across several of its civil works activities for water resources project planning. In addition to the Corps's internal planning procedures, risk analysis also influences the levee certification process that the Corps conducts jointly with FEMA.

As part of the Water Resources Development Act of 1996 (WRDA 96), the National Academy of Sciences (now part of the National Academies) was requested to conduct a study of the Corps's risk analysis methodology in its flood damage reduction studies. The 104th Congress of the United States passed Public Law 104-303 on October 12, 1996, which states in Section 202h the following:

> The Secretary (Army) shall enter into an agreement with the National Academy of Sciences to conduct a study of the Corps of Engineers' use of risk-based analysis for the evaluation of hydrology, hydraulics, and economics in flood damage reduction studies. The study shall include—
>
> a) an evaluation of the impact of risk-based analysis on project formulation, project economic justification, and minimum engineering and safety standards; and
> b) a review of studies conducted using risk-based analysis to

determine—
> i)   the scientific validity of applying risk-based analysis in these studies; and
> ii) the impact of using risk-based analysis as it relates to current policy and procedures of the Corps of Engineers.

A committee of the National Research Council's (NRC) Water Science and Technology Board (WSTB) was convened in late 1998 to address this charge, completing its study in May 2000.

THE CORPS'S WATER RESOURCES
PROJECT PLANNING PROCEDURES

The Corps's use of risk analysis techniques is applied to certain hydrologic, hydraulic, geotechnical, and economic aspects of Corps planning decisions for flood damage reduction studies. Development, refinement, and application of these risk analysis techniques occur within a larger context of Corps planning procedures, federal planning guidelines, and other considerations. This section describes Corps planning and decision making in its water resources projects and draws partly from a recent National Research Council study (NRC, 1999a) of the Corps's planning procedures.

The Corps has been conducting studies and constructing projects to manage the nation's waterways for nearly 200 years. In addition to its flood damage reduction responsibilities, the Corps enhances and maintains navigability on the nation's rivers (some Corps dams also generate hydroelectric power). In its civil works program for water resources development, the Corps is also involved in harbor improvements, hurricane damage prevention, and beach and shoreline protection. The Corps is also becoming more involved in ecosystem restoration; for example, the Corps plays a key role in the current effort to restore aquatic ecosystems in Florida's Everglades.

Several pieces of federal legislation and internal Corps planning documents guide Corps water resources project planning. One important document is the federal *Economic and Environmental Principles and Guidelines for Water and Related Land Resources Implementation Studies* (USWRC, 1983), familiarly known as the *Principles and Guidelines,* or simply the *P&G.* The Corps's *Planning Guidance Notebook* (USACE, 2000) is another key document, and contains advice on implementing the *Principles and Guidelines* within Corps planning studies. Corps planning procedures are further governed by the *Digest of Water Resources Policies and Authorities* (USACE, 1999a), guidance letters, and the Corps's own engineering regulations, engineering circulars, and engineering manuals (EM). The Corps is also obliged to conduct its studies pursuant to federal and state legislation and regulations, such as the National Environmental Policy Act of 1969.

From *Principles and Standards* to *Principles and Guidelines*

The predecessor to the *P&G* was the *Principles and Standards* (*P&S*). Adopted as regulations in 1973, the *Principles and Standards* is formally known as *Water and Related Land Resources: Establishment of Principles and Standards for Planning* (USWRC, 1973). This document defined four sets of objectives for U.S. federal water resources project plans: (1) national economic development (NED), (2) environmental quality (EQ), (3) regional economic development (RED), and (4) other social effects (OSE).

According to the *Principles and Standards*, water resources projects were to be evaluated primarily by their effects on the first two objectives—national economic development and environmental quality. The National Economic Development alternative was the water development plan that maximized economic development benefits for the nation, while the Environmental Quality alternative was the plan designed to minimize negative environmental impacts. The two secondary objectives, regional economic development and other social effects, could be assessed but were not required for all projects. The *Principles and Standards* also required that nonstructural alternatives be considered, that environmental mitigation measures be evaluated, and that a water conservation plan be included among the alternatives.

Although the *Principles and Standards* represented "the most detailed attempt to insure a broad range of choice in United States water planning" (Wescoat, 1986), they were repealed in 1982 and replaced by the *Principles and Guidelines* in 1983. The *Principles and Guidelines* represented an important departure from the *Principles and Standards* in at least two ways. First, with the change from "standards" to "guidelines," the planning document became merely recommended guidance rather than a requirement, thereby losing much of its regulatory force. Second, the *Principles and Guidelines* required the development of only one water project alternative, the National Economic Development alternative. According to the *P&G*, this alternative is to "contribute to the national economic development consistent with protecting the Nation's environment, pursuant to national environmental statutes, applicable executive orders, and other Federal planning requirements" (USWRC, 1983). Box 1.1 provides further discussion of the *Principles and Guidelines* and the National Economic Development alternative, especially as that alternative relates to flood damage reduction studies.

The *Principles and Standards* and the *Principles and Guidelines* were both established by the U.S. Water Resources Council (WRC[1]). The Water Resources Council was an executive-level agency created in 1965 to help coordinate and centralize federal-level water resource planning. The Water Resources Council was zero-funded in 1981 and therefore effectively terminated.

The *Principles and Guidelines* describe a six-step planning process:

1. specify problems and opportunities,
2. inventory and forecast conditions,
3. formulate alternative plans,
4. evaluate effects of alternative plans,
5. compare alternative plans,
6. select recommended plan.

The Corps uses these steps in its water resources planning, although they are not necessarily applied in this sequence. Formulation of plan alternatives, for example, may occur at various stages throughout a planning study.

The Corps conducts its water resources project planning studies in two separate phases: a reconnaissance phase and a feasibility phase. An idealized time line showing these two study phases is depicted in Figure 1.1. (See NRC 1999a for a more detailed discussion of the Corps's water resources planning procedures).

The reconnaissance phase of a Corps planning study is conducted to determine whether a water or related land resources problem warrants federal participation in feasibility studies and to define the federal interest, consistent with U.S. Army policies (USACE, 1999a). The reconnaissance phase ends with a recommendation to either terminate or

---

[1] The Water Resources Council originally consisted of the secretaries of Agriculture, the Army, Commerce, Energy, Housing and Urban Development, the Interior, and Transportation and (starting in 1970) the administrator of the Environmental Protection Agency.

BOX 1.1
The Principles and Guidelines: National Economic Development and Flood Damage Reduction Studies

The *Principles and Guidelines* (or *P&G*) is a key guidance document for all Corps of Engineers water resources project planning studies, including flood damage reduction studies, and has been referred to as the Corps's "philosophical source document" (Yoe and Orth, 1996). The *Principles and Guidelines* has its roots in the 1970 Flood Control Act, in which Congress identified four equal national development objectives for water resources project planning: 1) national economic development, 2) regional economic development, 3) environmental quality, and 4) social well-being.

In 1971 the Water Resources Council (WRC) issued the **Proposed Principles and Standards for Planning Water and Related Land Resources**. Major changes from the 1970 Flood Control Act were that social well-being was dropped as an objective, and that a plan maximizing contributions to national economic development would be required. After two years of review, the WRC in 1973 published the *Principles and Standards for Planning Water and Related Land Resources* in the Federal Register. The *Principles and Standards* placed environmental concerns on an equal basis with national economic development.

In 1983 the *Principles and Standards* were replaced by the *Economic and Environmental Principles and Guidelines for Water and Related Land Resources Implementation Studies (P&G)*. A major change from the *Principles and Standards* was that environmental quality was dropped as a planning objective, leaving national economic development as the sole required water resources project plan.

The "principles" comprise a two-page statement (Appendix C) that ensures consistent planning by the federal agencies that conduct water resources planning studies (the Corps, the Bureau of Reclamation, the Natural Resources Conservation Service, and the Tennessee Valley Authority). The "guidelines" consist of a) standards, b) National Economic Development (NED) procedures, and c) environmental quality evaluation procedures.

Chapter 2 of the *Principles and Guidelines* describes the procedures for estimating benefits for a range of water resource project planning studies. The approach and philosophy to estimating benefits under the National Economic Development is to "estimate changes in national economic development that occur as a result of differences in project outputs with a plan, as opposed to national economic development without a plan" (Yoe and Orth, 1996). Chapter 2 of the *P&G* describes specific procedures for estimating National Economic Develop-

BOX 1.1 Continued

ment benefits for the following types of water resources planning studies: Municipal and Industrial (M&I) Water Supply; Agriculture; Urban Flood Damages; Hydropower; Inland Navigation; Deep-Draft Navigation; Recreation; Commercial Fishing; Other Direct Benefits; Unemployed or Underemployed Labor Resources; and NED Costs.

The economics analysis in Corps flood damage reduction studies is conducted using a discount rate that reduces future benefits so that they can properly be compared to present costs. All benefit-cost comparisons are made using annualized benefits and costs over an expected project life. The discount rate is used so that the annualized cost in any year reflects the actual timing of the years in which the benefits and costs actually occur (this is discussed in greater in detail in Chapter 5 of this report).

Section IV of Chapter 2, "NED Benefit Evaluation Procedures: Urban Flood Damage," provides specific steps (see flow chart in Figure 1) for calculating benefits in a flood damage reduction study, as well as specific guidance on the types of benefits that can be included in calculating the NED alternative (see Table 1 below). Regarding the benefits in a flood damage reduction study allowed within the *P&G* planning framework, Section IV states "Benefits from plans for reducing flood hazards accrue primarily through the reduction in actual or potential damages associated with land use" (USWRC, 1983, p. 32). Section IV identifies three benefit categories: 1) inundation reduction benefit, 2) intensification benefit, and 3) location benefit. The Corps has applied these steps (Figure 1) and benefit estimation guidelines in its flood damage reduction studies for nearly twenty years without significant modification.

While the *Principles and Guidelines* represented the state-of-the-art in water resources planning when they were enacted in 1983, there have since been significant advances in economic and other analytical techniques, advances in aquatic biology, and shifts in public values related to water and related resources. A previous National Research Council (NRC) committee, formed in part to discuss possible revisions to the *P&G*, recommended "that the federal *Principles and Guidelines* be thoroughly reviewed and modified to incorporate contemporary analytical techniques and changes in public values and federal agency programs" (NRC, 1999a, p. 4). That NRC committee also reviewed the guidance offered by the *P&G* for federal flood damage reduction programs, and concluded: "the *P&G* do not allow for the benefits of primary flood damages avoided to be claimed as benefits in all nonstructural projects. The committee recommends that the benefits of flood damages avoided be included in the benefit-cost analysis of all flood

BOX 1.1 Continued

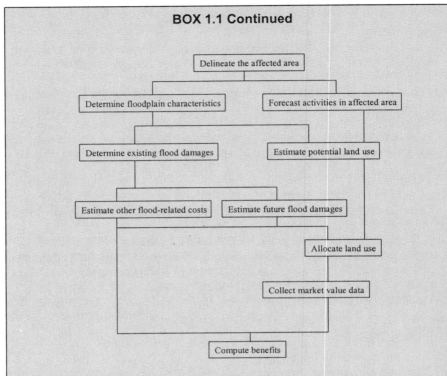

FIGURE 1 The Corps uses these ten steps to compute benefits. They are designed to determine land use and to relate use to the flood hazard from a NED perspective.

damage reduction projects—including nonstructural projects—and that these benefits be calculated in a uniform and consistent fashion" (NRC, 1999a, p. 8).

There appears to be some movement toward broadening water resources project planning as defined within the *Principles and Guidelines*. For example, the Corps's "Challenge 21" program (part of the 1999 Water Resources Development Act) seeks to include conservation and restoration of natural ecological functions as benefits of flood damage reduction studies. And just as this report was going to press, President Clinton issued a draft memorandum to the departments of Army, Interior, Agriculture, Commerce, and to the Council on Environmental Quality, to EPA, and to FEMA, in which he noted that he was directing the Secretary of the Army to develop proposed revisions to the 1983 *Principles and Guidelines*.

BOX 1.1 Continued

TABLE 1 Benefits Claimed/Not Claimed in Flood Damage Reduction Studies

Type of Benefit (And Step)	Structural	Floodproofing	Evacuation
Inundation			
Incidental flood damages (step 6)	Claimable	Claimable	Claimable
Primary flood damages (step 6)	Claimable	Claimable	Not claimable
Floodproofing costs reduced (step 7)	Claimable	Not claimable	Not claimable
Reduction in insurance overhead (step 7)	Claimable	Claimable	Claimable
Restoration of land value (step 9)	Claimable	Claimable	Not claimable
Intensification (steps 7 and 9)	Claimable	Claimable	Not claimable
Location			
Difference in use (step 9)	Claimable	Claimable	Not claimable
New use (step 9)	Not claimable	Not claimable	Claimable
Encumbered title (step 9)	Not claimable	Not claimable	Claimable
Open space (step 9)	Not claimable	Not claimable	Claimable

---

continue the study. This phase is to be completed in no more than 12 months, is to cost no more than $100,000, and is fully funded by the federal government. The reconnaissance phase of a Corps planning study is also used to create a project study plan, which describes the arrangements between the Corps and the project cosponsor for tasks beyond the reconnaissance study.

At the end of the reconnaissance stage, the Corps and the project cosponsor sign a feasibility cost sharing arrangement describing the details of project cost-sharing. Many terms of the feasibility cost-sharing arrangement are nonnegotiable, having been specified in legislation (e.g., the Water Resources Development Act of 1986).

Feasibility study costs are divided equally between the federal government and the project cosponsor. Risk analyses are conducted in the feasibility phase of a Corps flood damage reduction study. Alternative plans are identified at the beginning of the planning process and these

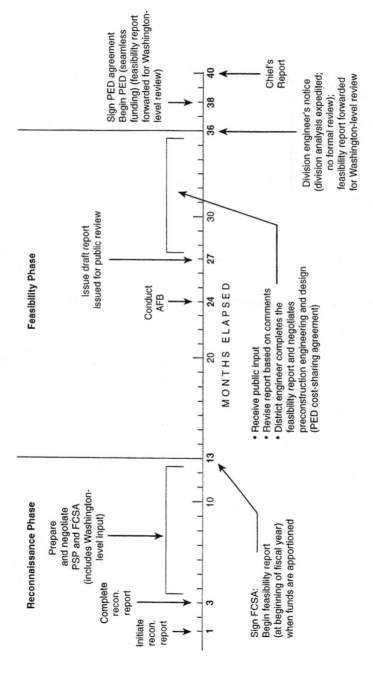

FIGURE 1.1 Recommended Corps planning study timeline. SOURCE: NRC (1999).

plans are screened and refined in subsequent iterations (USACE, 1999a, p. 5-4). As the *Principles and Guidelines* state, however, "A plan recommending Federal action is to be the alternative plan with the greatest net economic benefit consistent with protecting the Nation's environment (the NED plan), unless the Secretary of a department or head of an independent agency grants an exception to this rule" (USWRC, 1983).

The National Economic Development alternative does not necessarily represent the Corps's or the local sponsor's preferred alternative. The Corps will construct water resources projects in accord with a project sponsor's wishes and will share the costs of that project. Corps water resources project cost-sharing guidelines are specified in the Water Resources Development Act of 1986 (WRDA 86) and are modified in WRDA 96. But if a local sponsor desires a project that is larger or more costly than the National Economic Development alternative, then the local sponsor is responsible for at least a portion of the extra cost. For example, a local project sponsor is responsible for 100 percent of the extra costs of levees built higher than the elevation of the National Economic Development levee alternative.

The *Principles and Guidelines* requirement that the Corps select the alternative that maximizes net economic benefits to the nation has important implications for risk analysis applications and the construction of Corps levees. In a Corps flood damage reduction study, levee height is determined according to the National Economic Development criterion (i.e., based on prescribed benefit calculation procedures), rather than according to a levee's ability to withstand a flood of a given magnitude. As the Corps's *Digest of Water Resources Policies and Authorities* states, "There is no minimum level of performance or reliability required for Corps projects; therefore, any project increments beyond the NED plan represent explicit risk management options" (USACE, 1999a).

These issues can be problematic when, for example, a community requests the Corps to construct a levee for protection against an extreme flood, such as a 200-year flood (e.g., the flood with a probability of 1/200 of occurring in any given year). The Corps will pay the federal share of the National Economic Development levee (a minimum of 50 percent and a maximum of 65 percent of levee construction, pursuant to WRDA 86 and WRDA 96 cost-sharing guidelines); the local community, however, must bear the additional cost of constructing a levee higher than the level designated in the National Economic Development alternative. For instance, assume the Corps's National Economic Development alternative calls for a levee that provides protection only up to the 85-year flood. In this case the local sponsor would be required to pay

100 percent of the additional cost of raising the levee to protect against floods that exceed the 85-year flood. Such costs can be significant even when the additional levee height appears to be small. Engineers remind us that levees are raised from the bottom, not from the top.

Corps of Engineers levees figure prominently in the National Flood Insurance Program, which is conducted under the authority of FEMA. The following section describes the roles played by FEMA and other federal agencies in flood hazard mitigation, response, and recovery activities, and Chapter 7 discusses the Corps–FEMA levee certification program in more detail.

U.S. FEDERAL FLOOD PREPAREDNESS, MITIGATION, AND RESPONSE ACTIVITIES

Just as land use planning in the United States is not a federal responsibility, neither is comprehensive land use planning in the nation's floodplains. Nonetheless, federal agencies conduct extensive flood hazard reduction programs. Despite an impressive array of activities, however, no one agency coordinates these programs and there are no comprehensive floodplain management plans at the federal level. Although it is a key federal agency involved in flood hazard management, the Corps's flood damage reduction activities are but a part of a larger effort—which includes other federal, state, tribal, and local governments—toward managing flood risks. Box 1.2 summarizes flood-related activities conducted by other federal agencies.

The Corps of Engineers today uses the term "flood damage reduction," as opposed to flood control. This is consistent with recognition that no flood damage reduction program can provide complete protection against all floods. As demonstrated in the Mississippi River flooding of 1993 and in the extreme flooding in eastern North Carolina in 1999, there are classes of floods that exceed most, if not all, human experience and simply cannot be fully controlled by reasonable engineering structures. Thus, despite the best efforts of several federal, state, tribal, and local agencies and organizations, it bears repeating that no program or set of engineering structures can provide absolute control of all floods.

The Corps's flood-related programs have historically been in the realm of "structural measures": dams, reservoirs, levees, walls, diversion channels, bridge modifications, channel alterations, pumping, land treatment, and related structures intended to modify the flow of flood waters through storage or diversion. The Corps has also implemented "nonstructural" flood damage reduction projects. Section 212 of the Water Resources Development Act of 1999, for example, authorized the Corps's "Challenge 21" initiative. This legislation calls for the Corps (in cooperation with FEMA) to "undertake a program for the purpose of conducting projects to reduce flood hazards and restore the natural functions and values of rivers throughout the nation" (H.R. document 106-298, 1999). It further states that "the studies and projects shall emphasize, to the maximum extent practicable and appropriate, nonstructural approaches to preventing or reducing flood damages."

The "Galloway Report"

A comprehensive review of U.S. floodplain management activities was conducted by a special Interagency Floodplain Management Review Committee following the tremendous Mississippi River flooding in the summer of 1993. The committee was headed by U.S. Army Brigadier General Dr. Gerald Galloway; its report is thus more familiarly known as the "Galloway Report."

The report pointed out the lack of inter-agency coordination in floodplain management, identifying this fragmentation as one of the nation's major flood management problems: "the division of responsibilities for floodplain management among federal, state, tribal, and local governments needs clear definition. Currently, attention to floodplain management varies widely among and within federal, state, tribal, and local governments" (IFMRC, 1994, p. vii).

The Galloway Report emphasized that floodplain management is a responsibility that must be shared among all levels of government. To help promote better coordination, the report recommended the following actions:

• The president should enact a Floodplain Management Act which establishes a national model for floodplain management, clearly delineates federal, state, tribal, and local responsibilities, provides fiscal support for state and local floodplain management activities, and recognizes states as the nation's principal floodplain managers;
• Issue an Executive Order clearly defining the responsibility of federal agencies to exercise sound judgement in floodplain activities; and
• Activate the Water Resources Council to coordinate federal and federal-state-tribal activities in water resources; as appropriate, reestablish basin commissions to provide a forum for federal-state-tribal coordi-

BOX 1.2
U.S. Federal Flood Hazard and Floodplain Management

Several federal agencies other than the Corps of Engineers and FEMA conduct flood damage mitigation, response, and recovery activities. This box does not describe all such programs but indicates the variety of federal-level activities.

The *U.S. Department of Agriculture* (USDA) provides emergency loans to farmers impacted by floods. The loans cover losses to crops, livestock, and farm buildings and machinery. The USDA's Natural Resource Conservation Service (NRCS) operates an Emergency Watershed Protection Program, which conducts local-level damage assessment and recovery planning (sometimes in cooperation with FEMA). The NRCS also provides technical assistance to local sponsors in a variety of watershed protection and conservation programs, some of which aim to reduce the magnitude of floods. The USDA's Farm Service Bureau oversees the Flood Risk Reduction Program, in which farmers contract to receive USDA payments on flood-prone lands in return for foregoing certain USDA program benefits. The USDA's Risk Management Agency sponsors a crop insurance program; about 22% of crop losses in the U.S. are caused by "excess moisture" (USDA, undated).

The *Department of Health and Human Services* (HHS) has federal responsibility for assisting citizens with health and medical problems during floods and other emergencies. The HHS Office of Emergency Preparedness (OEP) coordinates federal health and medical response and recovery activities for HHS, working with other federal agencies and the private sector. HHS is the primary agency for health, medical and health-related social services under the Federal Response Plan, which provides for medical, mental health and other human services to disaster victims.

The *Department of Housing and Urban Development* (HUD) provides Community Development Block Grants to communities damaged by floods and other disasters. HUD's Disaster Recovery Initiative (DRI) Grants provide financial assistance to low- and moderate-income families that may have extreme difficulties in rebuilding after floods and other disasters. HUD employees also provide technical expertise in housing and construction to local officials as they develop strategies to rebuild and renovate communities after floods.

The *Department of Transportation* (DOT) operates the Federal-Aid Highways Emergency Relief (ER) program within the Federal Highway Administration. The ER program provides states up to $100 million in emergency relief funding for highways damaged by floods or other natural disasters.

The *Federal Emergency Management Agency's* (FEMA) flood-related responsibilities revolve around flood mitigation, response, and recovery activities and its administration of the National Flood Insurance Program. Communities can participate in the National Flood Insurance Program if they agree to regulate future floodplain construction by assuring that future structures are built to safe standards. For participating communities, FEMA makes federal flood insurance policies available to property owners (flood insurance is not underwritten by private insurers). As of May, 2000, federal flood insurance through the NFIP was available in over 19,000 communities in the U.S. and Puerto Rico. FEMA also conducts a variety of other programs related to flood hazard mitigation and response activities. For example, FEMA's Flood Mitigation Assistance (FMA) provides funding to assist states and communities in implementing measures to reduce or eliminate the long-term risk of flood damage to buildings, manufactured homes, and other structures insurable under the National Flood Insurance Program.

The *National Aeronautics and Space Administration* (NASA) is involved in flood hazard reduction through an agency-wide effort, the "Natural Disaster Reduction Program: Space Access and Technology, Management Systems and Facilities, Life Sciences, and Space Communication." Within this program, NASA's Solid Earth and Natural Hazards Program aims to improve local and regional flood forecasts by incorporating satellite-derived data on parameters such as topography, land cover, and soil moisture into watershed modeling research. NASA also conducts research on the consequences of inter-annual climate variability for rainfall and storm patterns.

The *National Weather Service* (NWS) sponsors the Hydrologic Information Center (HIC), which prepares national summaries of hydrologic conditions, including river conditions with emphasis on extreme events such as floods. In late winter and early spring, the HIC issues national outlooks for flood conditions based on data from its river forecast centers, weather forecast offices, and national Centers. The NWS also provides data on fatalities and some loss estimates in floods.

The *Small Business Adminstration* (SBA) operates a Disaster Loan Program that offers financial assistance through low-interest (4-8%) loans for renters, and home and business owners, who have suffered damages from a flood or other natural disaster.

The *U.S. Geological Survey* (USGS) operates a nation-wide system of stream gages that provides data used in making flood forecasts. The USGS is currently helping FEMA update its flood inundation maps through the use of geographic information system (GIS) technology, high-accuracy digital elevation data and models, and existing hydraulic models. The USGS also compiles and reports information on major floods across the nation.

nation on regional issues.

The report also recommended goals for floodplain management: "Establish, as goals for the future, the reduction of the vulnerability of the nation to the dangers and damages that result from floods and the concurrent and integrated preservation and enhancement of the natural resources and functions of floodplains. Such an approach seeks to avoid unwise use of the floodplain, to minimize vulnerability when floodplains must be used, and to mitigate damages when they do occur" (IFMRC, 1994, p. viii).

This chapter has described the planning, policy, and inter-agency context in which the Corps executes its flood damage reduction studies and plans. Although the Corps plays several important roles, they clearly are but a part of larger efforts in flood damage reduction. And, as the Galloway Report stressed, the different components of that effort require much better coordination than they have had to date. These inter-agency relations are further examined in Chapter 7, which describes coordination between the Corps and FEMA in a federal levee certification program within the National Flood Insurance Program.

2

Decision Making and Communication Issues

Floodplains constitute about 7 percent of the U.S. land area (Kusler and Larson, 1993) and represent a valuable national resource. Floodplains store flood waters during high flows (helping recharge groundwater supplies in the process); are a source of biological productivity and diversity; and are used for many human activities, including agriculture, grazing, parks and recreation, transportation, housing, and commercial development. However, because most of these activities preclude water storage during high flows, they need to be properly managed.

Individuals, local authorities, state governments, and several federal agencies make decisions about floodplain management. Even though it is in the national interest to do so, coordinating these decisions is exceedingly difficult, as different decision-making authorities have different interests and mandates. Furthermore, the overall goals of U.S. floodplain management are neither clearly specified nor well organized. Floodplain management decisions thus tend to be fragmented, as pointed out in the Galloway Report and elsewhere.

The Corps's risk analysis techniques and flood damage reduction studies will produce their greatest benefits if these techniques and studies are executed within a comprehensive planning paradigm and framework designed to make the best social, economic, and environmental uses of the nation's floodplain resources. Even the best analytical techniques will fall short of their potential contributions if flood damage reduction project goals are not consistent with public values, which can often be better determined through public participation and communication. This chapter reviews goals, multiple project objectives and trade-offs, decision making, and communication in floodplain management.

THE GOAL OF FLOODPLAIN MANAGEMENT

Neither the U.S. Congress nor the Corps of Engineers has identified an explicit goal for management of the nation's floodplains. Perhaps the closest that any federal water planning document comes to identifying a goal for floodplain management is the *Principles and Guidelines* (USWRC, 1983), which states that flood damage reduction projects (and other federal water projects) are to "contribute to the national economic development, consistent with protecting the Nation's environment."

The committee believes that the goal for management of the nation's floodplains should be broader: *to use the land for the greatest social benefit,* accounting for the risks of flooding and steps that can be taken to reduce those risks. In contrast, a goal such as minimizing damages from floods necessitates removal of people and activities from the floodplain. Removing people and activities from the floodplain, however, forfeits the many benefits of floodplain use and may thus be economically and socially undesirable.

The goal of maximizing social benefit in floodplain management leads to a strategy that recognizes the availability of land not in danger of flooding, the probabilities and magnitudes of potential floods, the availability of insurance, and the costs of flood damage reduction structures to reduce damages should a flood occur. As geographer Gilbert White stated: "It is striking that in a century of evolving public policy the prevailing aim has been to minimize losses from floods and not to optimize the net social benefits from using floodplain resources . . . . In simplest terms, it is the contrast between 'loss reduction' and 'wise use' (White, 2000).

It is unclear whether this approach promotes the wisest use of the nation's floodplains. The issue is important not only in its historical context, but also because of the damages exacted by floods: in most years, floods cause more deaths and damages than any other natural phenomenon, and the damages from floods in the U.S. are increasing over time (Richards, 1999). The distribution, frequency, and intensity of extreme weather events are changing in ways that are difficult to understand and predict (Karl et al., 1996), and an increasing Gross Domestic Product and population make it natural to expect greater demand for land in the floodplain. This in turn means greater damages in future extreme floods. Increases in wealth and increases in population in the nation's floodplains put more property at risk from floods. The trend of increasing damages thus does not necessarily imply a failure of the nation's approach to flood management (cf. Lave et al., 1990; Pielke, 1999).

Floodplain structures should not be considered fixed and immutable. In some areas the probabilities of flooding are sufficiently large, and the ability to mitigate flooding so expensive, that structures in the floodplain should be removed and activities relocated. Such a floodplain could be devoted to the highest-value uses consistent with periodic flooding (e.g., parks and recreation areas). In contrast, some currently underused floodplains are becoming increasingly valuable because of increasing population and economic activity. If after paying the costs of the protection structures and accounting for the costs of periodic flooding, net social benefits remain, these floodplains should be developed.

Land use controls, zoning, and planning are other important factors that complicate floodplain management. As much of the land in the nation's floodplains is privately owned, decisions about uses of those lands lies beyond the direct responsibilities of the Corps and most water management agencies. On these lands, there are often few incentives that encourage proper flood planning and preparedness (e.g., devoting lands to outdoor recreation activities, or elevating buildings above the 100-year flood stage). There are often also inadequate regulations that limit or prohibit development in flood-prone areas.

In the United States, strong tensions often exist between regulation and zoning on the one hand, and individual property rights on the other. But some U.S. communities have enacted programs for the purchase of floodplain properties that have sustained repetitive flood damages. These buyout programs aim to move susceptible property and its inhabitants out of high-hazard areas, while zoning these areas for land uses such as golf courses and hiking trails (which also serve as stormwater retention basins). For example, results from a comprehensive flood hazard mitigation and zoning program in Tulsa, Oklahoma are impressive: Tulsa's flood insurance rates have dropped by 25 percent and are now the lowest in the nation. In 1992, Tulsa received the nation's highest rating in the National Flood Insurance Program's Community Rating System (National Wildlife Federation, 1998). The Tulsa experience demonstrates the importance of local-level planning and decisions in effective floodplain management.

MULTIPLE OBJECTIVES

Although the *Principles and Guidelines* focus on economic benefits, issues other than economic damages should figure in the design and function of flood damage reduction projects. These issues include water quality, recreation, ecological protection, biodiversity, the quality of life, and life itself. In addition to the planning requirements of the *Principles and Guidelines*, the Corps is required by the National Environmental Policy Act to identify environmental implications of its projects. The Endangered Species Act requires the Corps to give special attention to select species whose survival might be compromised by a project. Many aspects of flood damage reduction projects transcend strict National Economic Development concerns.

Making decisions within this multiobjective framework is challenging. In some cases a single alternative dominates across all flood damage reduction objectives, making for an easy decision. In practice, however, the situation is seldom so agreeable. If a dominant alternative is quickly identified, it is likely that not enough thought was given to identify alternatives that are either less expensive but provide less flood protection, or are more expensive but provide more flood protection. The flood damage reduction project that provides the most benefits on one of the objectives seldom provides most benefits on all of them. In most cases, there are a range of project alternatives, all of which contain a complex mix of benefits and costs that must be weighed against each other.

If all people agreed on the values to be assigned to all project consequences, and if all these values allowed translation into a single dimension (e.g., dollars), multiobjective decisionmaking would be easy. For example, if a dollar value could be unequivocally assigned to extinction of a species, another dollar value to increases in water pollution, and so forth, the decision would come down to choosing the proposal with the greatest monetary net benefit, where each dimension is measured in dollars. However, there generally is little agreement about, for example, the social and economic consequences of species extinction, of reduced salmon migration, or of frequent flooding, and less agreement about how to express such outcomes in monetary units. At its worst, the inability to reduce complex issues to a small number of variables makes the situation akin to an environmental impact statement in which hundreds or thousands of impacts are identified, but there is no way to compare the impacts with one another.

The committee does not wish the Corps to become mired in such a morass. However, to the extent possible, the Corps should account for important social consequences of each project alternative, such as lives at risk in the event of flooding, and important environmental consequences, such as loss of wetlands or biodiversity. The *Principles and Guidelines* mandate the Corps to adopt the National Economic Development (NED) alternative in its water resources project planning studies (the alternative with the largest net economic benefits to whomever they accrue). However, the range of benefits to be counted in flood damage reduction studies, as specified by the *Principles and Guidelines*, may be narrowly construed (see Box 1.1). For instance, a previous NRC committee charged to review the Corps's water project planning procedures concluded, "Today, ecological and social considerations are often of great importance in project planning and should not necessarily be considered secondary to the maximization of economic benefits. Strict adherence to the NED account may discourage consideration of innovative and non-structural approaches to water resources planning . . . . The notion of NED as formulated in 1983 may not fit contemporary planning and social realities" (NRC, 1999a, p. 4). That committee went on to recommend a comprehensive review and modification of the *Principles and Guidelines*.

To enhance social benefits of floodplain management, this committee recommends that the Corps account for (or, more properly, be guided by the *Principles and Guidelines* to account for) a broad range of social and environmental considerations in its flood damage reduction studies and projects. Environmental, health, safety, and other social considerations of flood damage reduction projects should be quantified to the extent possible and included in floodplain management decisions.

COMPARING PROJECT ALTERNATIVES

The specific purposes of using risk analysis for flood damage reduction studies are to define project objectives, create desirable alternatives, evaluate those alternatives, guide analytical efforts, and facilitate communication. The ultimate intent should be good decisions that maximize net social benefits.

Complex decisions like those involved in flood damage reduction studies can be analyzed with a two-part decision model. The first part of the model relates decision alternatives to possible consequences. The second part assesses the relative desirability of the possible consequences. The two parts of the model are combined to establish relative desirability among a set of alternatives. This ranking derives from (1) the likelihood that particular consequences will result from an alternative, and (2) the relative desirability of those consequences. The likelihoods of consequences are estimated using scientific reasoning from data, while the desirabilities are based on value judgments. Clearly, the approaches to quantifying likelihoods and desirabilities are different, as are the people who should be making the quantifications.

Sound and comprehensive floodplain management is not solely a technocratic process. The values that society places on its rivers, flood-plains, wetlands, and water resources should be central to comprehensive flood damage reduction studies. The foundation for specifying public values is a logical set of specific concerns. As these concerns define the scope of public values relevant to a decision, it is important to directly involve citizens or representatives of citizen groups.

A value model can be constructed in four steps (Keeney, 1992): (1) identify values appropriate for the problem being addressed, (2) define and structure specific objectives related to those values, (3) specify at-tributes (i.e., metrics) with which to measure each objective, and (4) specify trade-offs among objectives. A specific objective might be stated as, "minimize economic damages," or "avoid loss of natural habitat." The attributes associated with these objectives might be, "damage meas-ured in dollars," and "destruction of habitat measured in acres."

The objectives can typically be categorized into four types (Keeney et al., 1996): (1) fundamental objectives, the ends used to describe con-sequences that are of concern to the public; (2) means objectives, the objectives that affect eventual consequences but which are themselves important only for their influence on the fundamental objectives; (3) pro-cess objectives, those concerned with how a decision is made rather than what decision is made; and (4) organizational objectives, those influ-enced by the complete set of decisions made over time by the organiza-tion with responsibility for acting in the public interest. Table 2.1 gives examples of these different types of objectives in the context of flood damage reduction.

The final step is to articulate value trade-offs among the objectives. Value trade-offs indicate willingness to forgo the achievement of one objective in order to increase achievement on another objective. Value trade-offs can be determined directly from the public or its representa-tives, and an experienced analyst can facilitate the assessments.

Schemes for making value trade-offs have been developed by, for example, Fishburn (1970), Keeney and Raiffa (1976), von Winterfeldt and Edwards (1986). Multiattribute decisionmaking is used to indicate the superiority of one alternative versus another, even when there are multiple attributes that cannot be compared quantitatively. Valuing non-market attributes, such as injury, disease, or polluted air, is done by looking at related decisions that involve market values. Examples include willingness to take more dangerous jobs in return for greater compensation, the increased value of property that is less polluted, the greater safety of some cars, and the willingness to pay for private campgrounds that are less crowded or less polluted.

A previous NRC committee charged to review the strategic plan of the U.S. Department of Interior's Grand Canyon Monitoring and Research Center noted the value of trade-off analysis in water resources management decisions: "It should be recognized that adaptive management for the Grand Canyon ecosystem will require trade-offs among management objectives favored by different stakeholder groups. The committee recommends that the Adaptive Management Work Group begin to consider mechanisms for equitable weighting of competing interests . . . . The Center's revised Strategic Plan should include a strategy for scientific evaluation of management alternatives, both in terms of ecological outcomes and satisfaction of stakeholder groups" (NRC, 1999b, p. 9).

TABLE 2.1 Representative Objectives and Their Relationships for Setting Public Policy for Levee Design

Category	Specific objective
Fundamental objectives	Maximize net economic benefit
	Maximize public health and safety
	Minimize construction costs
	Minimize environmental impacts
	Minimize social disruption
	Promote equity and fairness
	Protect agriculture
Means objectives	Ensure quality control
	Prevent damages
	Promote conservation of resources
	Minimize accidents
	Minimize construction impacts
Process objectives	Communicate with all stakeholders
	Coordinate with other decisions
	Involve the public
	Use reliable and accurate information
Corps organizational objectives	Contribute to public trust
	Ensure public acceptance

Reasonable value trade-offs can help decisionmakers make good public policy decisions. For other problems, they allow for the justifiable elimination of some inferior alternatives, leaving a smaller set of better alternatives.

FLOODPLAIN MANAGEMENT ALTERNATIVES

The Corps's flood-related programs have traditionally focused on structures intended to modify flood flows through storage or by changing channel and floodplain hydraulics. These include dams and reservoirs, levees, walls, diversion channels, bridge modifications, channel alterations, pumping, and land treatment. The alternatives considered are typically combinations of possible structures that will reduce flood frequency and magnitude. However, water resources project planning studies may benefit by considering, both alone and in combination with structural alternatives, alternatives that manage possible consequences of floods just prior to or during a potentially serious flood. These alternatives include warning systems, evacuation plans, and flood triage.

One way to lessen potential loss of life and flood damages is to warn people in sufficient time so that they, and even some of their possessions, can be removed from harm's way. Even a short warning time can be sufficient to avert loss of life from floods (Brown and Graham, 1988; Paté-Cornell, 1984). For example, the National Weather Service reports the formation of tropical depressions that could develop into hurricanes. It gives periodic warnings as the storm develops and indicates whether it appears likely that the storm will strike land. As time passes the quality of information concerning the storm's intensity and path increases. People can interpret this information to take steps to protect their lives and property, even when it is not certain that a damaging storm will strike them. Periodic updates help individuals make decisions about when the risk is high enough to take actions such as evacuation, boarding up windows, and buying emergency food and water supplies. This warning system has developed to the point where even a major hurricane may cause few deaths and where property damages are significantly reduced.

When a large flow threatens the integrity of a flood damage reduction system, a variety of trade-offs must be faced. For example, levees protecting farmland could be breached to lower the chance that levees protecting an urban area will fail. Even levees protecting one urban area could be breached to prevent levee failure in another urban area, where failure could result in much more catastrophic damage or loss of life.

Such triage decisions are difficult, because they may involve subjecting one area to flooding in order to lessen the chance that another will be flooded. These difficult decisions are often easier to make if they are discussed before flood conditions are imminent.

RISK COMMUNICATION

Identifying sound, credible, and effective risk reduction priorities and solutions depends greatly on a well-informed public. The public should be knowledgeable about risk issues and should be given opportunities to express opinions and become involved in risk assessment and risk management activities. This involves risk communication: the effective understanding of risks and the transfer of risk information to the public, and the transfer of information from the public to decisionmakers.

Risk communication covers a range of activities directed at increasing the public's knowledge of risk issues and its participation in risk management. It includes, for example, public education about hazards and public hearings on risk management. Much risk communication research has been conducted since the early 1980s, when risk communication emerged as a distinct element of risk analysis (Fischhoff et al., 1981; Morgan et al., 1992).

The intent is to explain how the public perceives and interprets risks and to identify ways to improve the transfer of information to the public. A large fraction of the public is unfamiliar with the nature of the risks to which they are exposed. As the Galloway Report concluded: "As the Midwest Flood of 1993 has shown, people and property remain at risk, not only in the floodplains of the upper Mississippi River Basin but also throughout the nation. Many of those at risk neither fully understand the nature and potential consequences of that risk nor share fully in the fiscal implications of bearing that risk" (IFMRC, 1994, p. xxi).

Risk management decisions should not simply be made by technical experts and public officials and then imposed on, and justified to, the public after the fact. Risk communication involves a dialogue among interested parties—risk experts, policy makers, and affected citizens. It also involves the news media, as citizens often receive their information from the media. If the media do not report knowledgeably and accurately, constructive public involvement becomes more difficult.

The public's response to risk issues is complex because "the public" contains groups with different values and stakes. The risk assessment process should be opened to participation and scrutiny by affected stakeholders. This will increase the need to facilitate the public's ability to understand risk information and the ability of policy makers to understand public perception of risk. It is worth noting that public involvement including large numbers of stakeholder groups requires a significant commitment of time and effort from members of the public, as well as staff members from the agencies involved in flood damage reduction. With larger and more expensive projects, years of commitment may be required to help facilitate communications.

Decisions about appropriate floodplain management strategies differ from decisions about proper communication of those decisions. The goal of floodplain management should be to use the floodplain for the greatest social benefit. The goal of the communication decisions is to involve and inform the public and to have them understand what floodplain management decisions were made and why. These decisions are ideally addressed simultaneously, as it is important to establish two-way communication with interested parties in the course of developing and analyzing floodplain management plans.

Ideal communication decisions involve several steps:

    1.   identify the audiences that should be communicated with and involved in the decision process about floodplain management,

    2.   specify the objectives of communication, including the information people should provide and should receive, and what requires action and what those actions are,

    3.   create alternatives for communication that include oral and written communication (and possibly internet options) for experimenting with models, and

    4.   select the best combination of alternatives after appraisal.

The quality of communication is greatly enhanced when trust exists among the parties involved in the communication. This trust is built through an open process in floodplain management decision making and by involving stakeholders early in the process (which the Corps often does in many of its water resources project planning studies). If individuals feel they are involved in analyzing floodplain alternatives that affect them, they are more likely to understand and accept the implications of "their study."

The methods for analyzing the complexities of floodplain management are not simple to understand. This makes it difficult to communicate with citizens who are unfamiliar with scientific principles (e.g., hydrology, structural design) necessary to design floodplain management facilities. Indeed, few of the individuals involved in floodplain management understand all these principles well. It is thus a challenge to have individuals understand the full details of a flood damage reduction planning study.

It is important to use simple models to describe methodological ideas and the results of analysis rather than, for instance, models that focus on the mathematical and scientific concepts used in the analysis. Indeed, most of the public is more concerned about a specific application of a method than about the method itself. It is thus often easier to illustrate both the ideas of a method and the specific application together.

Once an analysis is completed, the critical factors that influenced the selection of recommendations can usually be identified. Simple models that illustrate these key ideas in simple situations that can be more easily understood might be especially useful. For instance, imagine a complex computer model involving more than 20 stages that analyzes alternative plans. The recommended plan depends strongly on the interaction of upstream and downstream management strategies. In this case it should be possible to build a simple two-or three-stage model with hypothetical, but realistic, information that reproduces the key interactions that are critical to the recommendations for the real problem. A simple situation model may also be very helpful in communicating key insights from a more complex analysis. This should enhance the likelihood that individuals will understand the interaction and hence be able to see how it is relevant to them. In illustrating these simple models, it is important to explicitly include all important assumptions and judgments, about both facts and values, that are relevant to the results.

Documentation of floodplain management studies is another critical aspect of communication. The standard for documentation is that an interested party should be able to understand everything that was done, why it was done, how it was done, and the range of implications. All assumptions and summaries of the value judgments and data used should be provided for anyone to examine. In the end, while they do not have to agree with all agency (the Corps and others) planning decisions, stakeholders should fully understand all the steps involved in the flood damage reduction study.

3

Risk Analysis Concepts and Terms

This chapter describes the Corps's progress in its risk analysis applications, the methods and terms the Corps employs in those applications, and provides recommendations regarding standardization of risk terminology and concepts.

Water resources project planning involves many types of uncertainties. Some of these relate to the natural environment, such as the variability of precipitation, stream flow, and river stage. Others relate to the performance of engineered systems, such as the reliability of levees, pumps, locks, and gates, or to variations in transit times of barges. Still others relate to the economic value of floodplain property, the probability distribution used to describe flood frequency, or the costs of alternative transportation modes.

The Corps has made significant strides in the use of risk analysis. In the committee's judgment, it would be advantageous for the Corps to consistently use terms describing uncertainty and to standardize risk and uncertainty concepts throughout its civil works programs. This would result in a clearer understanding of risk analysis issues among Corps personnel, most of whom work across program areas. It would also facilitate communication with other federal agencies, consultants, contractors, and the public.

UNCERTAINTY

The term "uncertainty" is used by different people to mean different things. A review of Corps documents describing risk-related planning activities suggests that a consistent set of terms can be readily developed, and this set should be agreeable to most of those involved in risk analyses (Table 3.1). This set of terms, as discussed below, is consistent with the usage of others, including Morgan and Henrion (1990), Haimes (1998), and the Corps's Institute for Water Resources (USACE, 1992a,b).

The term *uncertainty* is normally used to describe a lack of sureness about something or someone, ranging from just short of complete sureness to an almost complete lack of conviction about an outcome. Doubt, dubiety, skepticism, suspicion, and mistrust are common synonyms. Each synonym expresses an aspect of uncertainty that comes to play in risk analysis. Uncertainty with respect to natural phenomena means that an outcome is unknown or not established and is therefore in question. Uncertainty with respect to a belief means that a conclusion is not proven or is supported by questionable information. Uncertainty with respect to a course of action means that a plan is not determined or is undecided.

In many, but not all, situations a lack of sureness can be described by probability distributions. The definition of uncertainty found in the *Principles and Guidelines* is that uncertainty describes only situations wherein the lack of sureness is not describable by probabilities. This narrow definition is no longer commonly used. The term *uncertainty* should be used to describe situations without sureness, whether or not described by a probability distribution.

Generally speaking, uncertainty can be attributed to two sources: (1) the inherent variability of natural processes ("natural variability"), or (2) incomplete knowledge ("knowledge uncertainty"). These two sources arise for different reasons and are usually evaluated in different ways (Morgan and Henrion, 1990). Moser (1998) and a National Research Council committee (NRC, 1996) describe these two types of uncertainty as follows:

Natural variability—sometimes called aleatory uncertainty—deals with inherent variability in the physical world; by assumption, this "randomness" is irreducible. The word *aleatory* comes the Latin *alea*, meaning a die or gambling device. In the water resources context, uncertainties related to natural variability include things such as stream flow, assumed to be a random process in time, or soil properties, assumed to be random in space. Natural variability is also sometimes referred to as external, objective, random, or stochastic uncertainty.

Knowledge uncertainty—sometimes called epistemic uncertainty— deals with a lack of understanding of events and processes, or with a lack of data from which to draw inferences; by assumption, such lack of

TABLE 3.1 Alternative Terms from the Professional Literature Describing Categories of Uncertainties

Type of Variability	Terms from Literature
Natural Variability	Aleatory uncertainty
	External uncertainty
	Objective uncertainty
	Random uncertainty
	Stochastic uncertainty
Knowledge Uncertainty	Epistemic uncertainty
	Functional uncertainty
	Internal uncertainty
	Subjective uncertainty

knowledge is reducible with further information. The word *epistemic* is derived from the Greek "to know." Knowledge uncertainty is also sometimes referred to as functional, internal, or subjective uncertainty (see Box 3.1).

IACWD (1981) provides an example of the distinction between natural variability and knowledge uncertainty found in flood–frequency calculations, wherein the frequency curve (i.e., probability distribution) describes natural variability, and the error bounds about the curve (i.e., uncertainty in the parameters of the probability distribution) reflect knowledge uncertainty. Natural variability is presumed to be an uncertainty of the world, a natural or inherent randomness. Knowledge uncertainty, in contrast, is presumed to be an uncertainty of the mind, a function of models and data.

Although the distinction between natural variability and knowledge uncertainty is both convenient and important, it is at the same time hypothetical. The division of uncertainty into a component related to natural variability and a component related to knowledge uncertainty is attributable to the model developed by the analyst. Consider flood frequency. In the future—at least in principle—the sophistication of atmospheric models might improve sufficiently such that flood time series could be modeled and forecast with great accuracy. All the uncertainty currently ascribed to natural variation might become knowledge uncertainty in the modeling, and thus reflect incomplete knowledge rather than randomness. Modeling assumptions may cause "natural randomness" to become knowledge uncertainties, and vice versa.

In its risk analysis framework, the Corps should be clear about which

---

BOX 3.1
Evaluating Knowledge Uncertainties

There are several examples from civil engineering in which risk analyses have involved the assessment of natural variability and knowledge uncertainty. One area of considerable development is earthquake engineering. Here, the assessment of seismic hazards and performance of civil, mechanical and electrical systems has involved comprehensive probabilistic analyses in which full analyses of natural and knowledge uncertainties are conducted as part of seismic risk studies. These analyses are performed for critical facilities such as nuclear power plants, chemical weapon demilitarization facilities, insurance risk assessments, and lifeline systems (e.g., water, gas, communication, and transportation).

In probabilistic seismic hazard assessments, assessment of knowledge uncertainties has significantly matured in two areas (although seismic hazard analysis is fundamentally an earth science endeavor, probabilistic modeling of seismic hazards was initiated and has been largely advanced by civil engineers (e.g., Budnitz et al., 1997; Cornell, 1968; and Cornell and Vanmarcke, 1969)). These are the development of models for estimating the spatial and temporal rate of earthquake occurrences (seismic source characterization), and the prediction of earthquake ground motions. The characterization of the seismic source is particularly challenging. While considerable historic and instrumental data are available, estimates of the spatial and temporal rate of earthquake occurrences must rely on scientific evaluations and probabilistic assessments in which indirect evidence of the potential for future earthquake occurrences are gathered and evaluated. As part of these evaluations, formal elicitations are conducted with the earth scientist who must quantitatively evaluate knowledge uncertainties in modeling the location, magnitude and frequency of future earthquake occurrences.

---

variables it treats as natural variability, which it treats as knowledge uncertainty, and why and how it makes this distinction. Furthermore, the Corps should establish a risk analysis framework that permits quantification of each source of uncertainty and properly incorporates each uncertainty in the analysis. Differences in the effects of these sources of uncertainty on risk calculations can be large. For example, variations in stream flow, treated as natural variability, average out in a calculation from one year to the next (high flows in one year balance against low flows in another). In contrast, uncertainty in the mean annual flow parameter, treated as knowledge uncertainty, introduces a systematic effect into a calculation. If the mean flow is overestimated in one year, it is overestimated in every year of the calculation.

It is not always obvious which uncertainties in a risk analysis should be ascribed to natural variability and which should be ascribed to knowledge uncertainty. Although most engineers and planners are familiar with natural variability, they are often less familiar with knowledge uncertainty. To understand how knowledge uncertainty enters a risk analysis, one might think of the analysis as being built upon a mathematical model describing the behavior of the natural world. Mathematical relationships in this model include parameters that determine how output varies with input—for example, the stability of a levee as water rises behind it. In its simplest form, knowledge uncertainty can be thought of as comprising uncertainty in the appropriate parameter values for the model, combined with uncertainty in the model itself. Parameter uncertainty relates to the accuracy and precision with which parameters can be inferred from field data, judgment, and the technical literature. Model uncertainty relates to the degree to which a chosen model accurately represents reality.

Parameter uncertainty derives from statistical considerations and is usually described either by confidence intervals when using traditional (frequentist) statistical methods, or by probability distributions when using Bayesian statistical methods. Data uncertainties, which are the principal contributors to parameter uncertainty, include (1) measurement errors, (2) inconsistent or heterogeneous data sets, (3) data handling and transcription errors, and (4) nonrepresentative sampling caused by time, space, or financial limitations.

Model uncertainty can result from the use of surrogate variables, from excluded variables, and from approximations and the use of the incorrect mathematical expressions for representing the physical world. An NRC committee argued that model uncertainty should be addressed with sensitivity analysis (NRC, 1994); however, this view is not unanimously shared by the scientific community.

Another type of knowledge uncertainty might be called *decision model uncertainty*, which describes an inability to understand the objectives that society holds important or to understand how alternative projects or designs should be evaluated. Such uncertainty, for example, would include uncertainty in discount rates and the appropriate length of planning horizons.

The Corps's risk analysis approach in flood damage reduction studies is mandated in ER 1105-2-101, *Risk-Based Analysis for Evaluation of Hydrology/Hydraulics, Geotechnical Stability, and Economics in Flood*

Damage Reduction Studies (USACE, 1996a), and further discussed in EM 1110-2-1619, *Risk-Based Analysis for Flood Damage Reduction Studies* (USACE, 1996b). While this latter document provides a clear definition of parameter uncertainty and model uncertainty, neither document discusses fundamental differences between natural variability and knowledge uncertainty.

The distinction between natural variability and knowledge uncertainty is particularly important for flood damage calculations of expected annual damage (EAD) of the form found in the Corps's risk analysis procedure. Such calculations of expected annual damage lead to different numerical results depending upon which uncertainties—natural variability, knowledge uncertainty, or both—are included in the probabilistic averaging. In the Corps's method, expected annual damage is calculated by averaging natural variations among floods and in levee performance. Thus, the expected annual damage so calculated contains no contribution from knowledge uncertainty. To incorporate knowledge uncertainty, a probability distribution is specified over expected annual damage. This probability distribution over expected annual damage reflects the influence of parameter uncertainties in the flood–frequency distribution, stage–discharge curve, and stage–damage function. The expectation of expected annual damage itself reflects only natural variability, while the probability distribution of expected annual damage reflects only knowledge uncertainty. Due to nonlinearities in the calculations, this procedure of separately treating natural variability and the knowledge uncertainty can lead to different results compared to the approach of incorporating both types of uncertainty from the beginning.

An NRC committee that reviewed flood risk management in the American River (California) basin (NRC, 1995) recommended that the Corps be clearer about which variables it treats as natural variability in the computation, which it treats as knowledge uncertainty, and why it makes the choice it does.

CONSISTENCY ACROSS PROGRAM AREAS

Risk analysis is based upon (1) the magnitude and likelihood of consequences, (2) defined risk acceptance criteria, and (3) a balance between implementation costs and avoided costs (Moser, 1998). In addition, such analyses should provide insight and understanding of likely failure modes and of significant economic issues.

Corps documents use a variety of terms to describe what in this report is called *risk analysis* (Table 3.2). Among these are risk analysis, risk-based analysis, and risk and uncertainty analysis. All of these use probability to assess likelihoods of events occurring. The terms appear to be used interchangeably to describe efforts involving probabilistic analyses. Flood damage reduction studies use "risk-based analysis" and "risk and uncertainty." Rehabilitation studies have often used "risk-based analysis." Environmental and ecosystem restoration studies typically use "risk and uncertainty analysis."

"Risk analysis" is the more general term that includes risk assessment and risk management (NRC, 1983) and sometimes also includes hazard identification, risk characterization, and risk communication (NRC, 1994, 1996). The Corps should adopt "risk analysis" as the most general term. For Corps water resources project planning purposes, no distinction should be made between risk analysis, risk-based analysis, and risk and uncertainty analysis.

"Risk" is generally understood to describe the probability that some undesirable event occurs, and is sometimes used to describe the combination of that probability and the corresponding consequence of the event. The Corps measures risk by the probability that system operation is undesirable (e.g., the probability that a levee fails or that an ecosystem restoration project fails to meet a standard). The complement of risk is *reliability,* the probability that a system operates without failing. In an economic risk analysis, the consequences of undesirable performance are also computed (e.g., expected flood damage).

An important document in the Corps's rehabilitation program area is *Tools for Risk Based Economic Analysis* (USACE, 1999c). In describing what constitutes a risk analysis, this document presented only knowledge (parameter) uncertainty. The discussion neglects natural variability. This is noteworthy because natural variability was the only uncertainty

TABLE 3.2 Terms employed in Corps program areas

Program Area	Term Used	Term Sometimes Used
Risk analysis course	Risk analysis	
Rehabilitation	Risk-based analysis	
Flood damage reduction	Risk-based analysis	Risk and uncertainty analysis
Environmental restoration	Risk and uncertainty analysis	
Dam safety	Risk analysis	

included in the other risk analysis programs described in that publication (rehabilitation for hydropower and for locks, channel improvements in waterways, and dike maintenance).

Efforts to develop risk analysis for environmental and ecosystem restoration projects have proceeded carefully in clarifying terminology and the conceptualization of uncertainty (USACE, 1996c). The Corps refers to a taxonomy suggested by Morgan and Henrion (1990) for categorizing different kinds of quantities in modeling (USACE, 1996c, 1996d). That taxonomy is then used to categorize instead different types of uncertainty (USACE, 1996c). The Corps correctly applies the taxonomy to quantities (USACE, 1996d); however, much of the confusion is retained in another Corps document (USACE, 1996c).

The terminology and concepts that underlie the use of risk analyses across Corps program areas are not always well documented and not always consistently applied. There would be clear advantages to having a consistent, well-documented conceptual framework and a consistent set of terms to support those analyses. With relatively little effort, this situation can be improved by adopting a set of terms similar to those in Figure 3.1.

RISK ANALYSIS AND DECISION MAKING

Improved decision making is emphasized in the Corps's risk analysis literature, and there is widespread interest in how these tools can be used more effectively. Table 3.3 shows four civil works program areas using risk analysis, and the performance metrics important to each. The choice of decision criteria is generally related to risks to human welfare or to large economic losses.

Many flood damage reduction studies and projects implicitly include some risk to human life. Such risk is described by the annual exceedance probability. Yet the primary decision criterion employed by the Corps, as specified by the *Principles and Guidelines*, is national economic development (NED). Supplemental criteria are the conditional nonexceedance probability for various design events and the expected annual damages (EAD). The current analytical approach does not address the question of which uncertainties are the more important.

Environmental restoration projects typically do not focus upon loss of life or on reducing flood damages. However, such projects have inherently low reliability, because habitat suitability models are often poorly developed and investment levels tend to be modest (USACE,

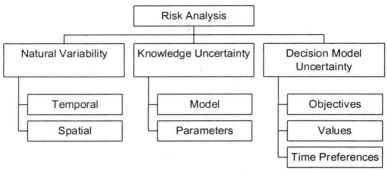

FIGURE 3.1 Taxonomy of uncertainties in risk analysis.

1997a). Adaptive management has been suggested as one approach for addressing uncertainties associated with ecological complexity (such as in Corps's environmental restoration efforts in the Florida Everglades and in Missouri River dam operations). It appears that formal risk analyses have been uncommon in environmental restoration studies.

Rehabilitation studies are typically not concerned with loss of life or even with large economic loss. As mandated by the *Principles and Guidelines*, national economic development serves as the primary decision criterion in rehabilitation studies. But the NED criterion can be supplemented by other performance metrics because, if expected costs of alternatives are essentially equal, a plan that minimized disruption is generally preferred. Risk analyses that explain the dynamics of a system and explain opportunities for interventions that improve system operation can be useful. At a minimum, risk analysis should identify which uncertainties are the most important.

Across these four areas, basic analyses have been formulated to compute primary and secondary criteria. To achieve the objective of using risk analyses to improve decision making, the remaining challenge is to compute other criteria that provide insight into system operation and into where cost-effective changes can be made to improve performance. It is similarly important to determine which uncertainties are important.

Does risk analysis aid decision making in flood damage reduction studies? The new risk and uncertainty analysis method developed for flood damage reduction studies is different from earlier methods in that it includes a wider range of parameter uncertainties in the stochastic Monte Carlo analysis that generates expected project damages. It is thus important that the distributions describing parameter uncertainty be appro-

TABLE 3.3 Performance Metrics in Corps Program Areas Using Risk Analysis

Program Area	Primary Criteria	Secondary Criteria	Threat to Human Life	Other Criteria
Rehabilitation	Expected costs (NED)		Low	• Annual failure rates • Hours/year for unscheduled outage • Trip transit time
Environmental Restoration[1]	Change in habitat units Expected costs Cost per habitat unit	Distribution of Habitat Units	Low	• Probability exceeds standard • Flexibility: ability to make adjustments
Dam Safety[2]	Annual loss of life Failure probability	Expected costs (NED)	High	• Cost per life saved
Flood Damage Reduction[3]	Expected damages (NED)	Probability of flooding (AEP)	Medium	• Conditional non-exceedance probabilities • Estimated Annual Damage quantiles

[1] The current criterion for ecosystem restoration projects is cost effectiveness or incremental cost analysis. The incremental analysis generally ignores uncertainty except when it is reflected in the value of specified characteristics of a site (e.g., water fluctuation, average temperature, annual maximum pH). The Corps (e.g., 1997a, p. 93–98) suggests use of the mean habitat unit change, minimum and maximum changes, and the distribution of the change as output from a risk and uncertainty study. The Corps is conducting additional research in this area.
[2] These are the anticipated performance metrics for this area.
[3] The failure probability is the probability of flooding, also called the annual exceedance probability (AEP). Expected damages are used in calculating the National Economic Development (NED) objective.

priate; otherwise, the expected annual damages criterion upon which projects are selected and justified will be distorted. This concerned both an NRC committee (NRC, 1995) and Stedinger (1997), who challenged the description of uncertainty in the parameters of the flood–frequency distribution.

The question has arisen whether adding parameter uncertainty to flood damage reduction calculations leads to potentially different project decisions or to greater insight into project performance. Some have noted that because the primary decision criterion is average expected annual damages, parameter uncertainty should have little impact. Stedinger (1997), for example, showed that with small sample sizes and high levels of protection, hydrologic parameter uncertainty can significantly increase expected damages; yet, Al-Futaisi and Stedinger (1999) found that adding hydrologic parameter uncertainty to the design process had little effect. Thus, while including uncertainty in economic analyses may impact performance indices, it may not impact the designs selected.

The influence of uncertainty on the expected value of performance criteria depends upon the nonlinearity of the models being used. A small uncertainty in the flow–stage relationship, or in the stage at which a levee fails, for example, can make a large difference in the reliability of a levee system and thus in project decisions. Anecdotal evidence from the American River (California) project suggests that risk analysis led to potentially significant changes in the operating rules for Folsom Reservoir, based on the capacity of a flood bypass and levee system downstream (M. Burnham, U.S. Army Corps of Engineers, personal communication, 1999). Further study is needed to assess how risk analysis can best be used in making project decisions for flood hazard damage reduction.

4

Risk Analysis Techniques

Multidisciplinary factors contribute to the risk of flooding in riverine systems. Yet levee safety planning in the United States has traditionally been conducted along disciplinary lines. For instance, in hydrology, levees are evaluated for their ability to withstand the flood of a given magnitude, usually defined by a return period. In geotechnical engineering, levees are assessed for their stability and potential for failure by seepage through the embankment. This disciplinary approach has inhibited the development of quantitative procedures that evaluate the total risk of flooding, reflecting possible contributions of various operational, hydrologic, hydraulic, and geotechnical factors and how they might act individually and jointly.

Among the various sources contributing to flood risk, only the flood–frequency component has traditionally been considered probabilistically. This component is indeed a major factor contributing to the flood risk, often accounting for more than 50 percent of the failure risk. But several other factors are significant and should be accounted for quantitatively. In addition, even for the flood–frequency component, conventional evaluation procedures may be incomplete.

Many factors contribute to flooding of protected areas, depending upon the hydrology and hydraulics of different riverine systems. Before starting a flood damage reduction study, it is important to differentiate between significant and insignificant factors. The following list addresses factors that are often important in determining flood risks.

Hydrologic factors — flood frequency and volume and time distribution of the flood along the stream, which in turn depend on snow melt and/or rainfall characteristics, rainfall–runoff relationships of the watershed, and the characteristics of the stream network. Rainfall factors include spatial and temporal distributions of the precipitation, the sample representativeness, accuracy and adequacy of the rainfall data, and the methods of analysis or simulation. Likewise, there are uncertainties in the representativeness, accuracy, and adequacy of the flood data in both space and time, and in the methods used to analyze these uncertainties. Watershed-stream factors include storage in lakes, reservoirs, and wetlands. There are also uncertainties in soil moisture, rain interception, and changing land uses.

Hydraulic factors — the nature of flood propagation in the channel and the equations and methods to simulate the flood propagation, which in turn depend on channel geometry, the roughness and slope of the channel bed, and the nature of the floodplain. Also included are the effects of hydraulic structures in the watershed, such as dams and spillways, levees, locks, weirs, sluices, gates, valves, bridges, intakes, and other diversion structures; also included are effects of sediment in the river, including erosion, scour, and deposition along the channel. Effects of wind and waves should also be considered.

Structural and geotechnical factors — geologic properties of the foundation, seepage through and cutoff beneath levees, internal erosion or piping of levee materials, strength instabilities in embankments or the subsurface, deep seepage failure away from the levee, and other soil mechanics issues.

Seismic factors (on dams and levees) — frequency and magnitude of earthquakes, fault and tectonic characteristics, earthquake-induced ground motion at the dam or levee site and liquefaction of foundation soil, and flooding probability associated with earthquake-induced dam or levee failure.

Materials and construction factors — type and quality of materials used for dams and levees, thermal and moisture variations affecting dam or levee quality during its service period and during its construction, and construction quality control.

Other geophysical factors — ice action in the river and on dams, levees or other structures, flash flooding from failure of dams, levees, or other facilities; thunder/lightning destruction; and tornado and other weather-related impacts.

Operational and maintenance factors — operational procedures on water diversion and release prior to and during flooding; operational procedures when an incident occurs; safety inspections of the river system; regulations on boat traffic and fishing during flooding; repair and maintenance rules; grazing and other land uses; and vegetation cover and type.

CORPS FRAMEWORK

The notion of using risk analysis to study the magnitude of floods is not new. Indeed, the relationship between the magnitude of a flood and its likely return period was established years ago by Gumbel (Gumbel, 1941), who drew on statistical theory developed during the 1920s concerning the distributions of extreme events. Standardized procedures for determining flood–frequency curves were defined by the Corps of Engineers during the 1950s (Beard, 1962). In the United States, passage of the National Flood Insurance Act of 1968, which created the National Flood Insurance Program (NFIP), also led to a period of comprehensive study of methods for determining flood frequency curves. This culminated in the publication in 1981 of the widely-used *Bulletin 17B* (IACWD, 1981). A similar comprehensive study of flood frequency was made in the United Kingdom at about the same time (NERC, 1975; Institute of Hydrology, 1999). Although the methods that evolved in the U.S. and U.K. studies are different, they have become standards adhered to since that time and are widely emulated in other countries.

A formal risk analysis includes seven phases. First, the level of unacceptable flooding performance is defined to allow a probabilistic failure analysis. Second, a method is identified that can be used to combine the different processes or events that lead to unacceptable performance. Third, the parameters involved in each of the processes or events is identified. Fourth, uncertainty analysis is performed for each of the parameters. Fifth, the component parameter uncertainties are combined to yield a system failure probability. Sixth, an economic damage function of flooding is determined along with associated uncertainty. Finally, the failure probability and damage function are combined to yield expected annual damage. This analysis is performed for each protection alternative considered.

The first five phases—finding the system failure probability, without considering the consequences of failure—are referred to as reliability analysis. A prerequisite of a successful reliability analysis is a comprehensive understanding of the problem and of the significant parameters involved. More than seventy years ago, industrial engineers applied some reliability techniques for quality control of manufactured products. In response to high failure rates and damages of military airborne and electronic equipment during World War II, the U.S. Joint Army–Navy Committees on Parts Standards and on Vacuum Tube Development were established in June 1943 to improve military equipment reliability. Carhart (1953) produced an early state-of-the-art report on reliability engineering. In November 1953 the U.S. Department of Defense set up the Advisory Group on the Reliability of Electronic Equipment to monitor and promote reliability evaluation and analysis. Textbooks on reliability engineering started to appear in the early 1960s (Bazovsky, 1961; Calabro, 1962). Concerns regarding the safety of nuclear power plants and the reliability of space vehicles further accelerated the development of this topic. In civil engineering, structural engineers have made considerable advances in understanding the risks that earthquakes and high winds pose to structures. Progress has also been made in geotechnical and water resources engineering (Yen and Tung, 1993).

NATURAL VARIABILITY AND IMPERFECT KNOWLEDGE

A risk analysis of flood hazards needs to address uncertainties associated with natural variability, engineering or economic models, and statistical relationships. Figure 4.1 provides a conceptual model (also known as an event tree) for describing the transformation of hydrologic risk, to variation in reservoir operations, to river-stage-to-reservoir-outflow relationship, to levee reliability, and finally to estimates of economic damages should a levee fail. This figure essentially represents the current Corps framework.

Some of the relationships in Figure 4.1 relate to natural variability. These relationships might be called *random* or *stochastic* because they are treated as random processes over time or space. For example, possible values of annual flood flows are treated as random events in time, and levee failures related to geotechnical weaknesses are treated, in part, as random events over space.

Other relationships in Figure 4.1 relate to engineering calculations or functional rules. These relationships might be called *deterministic* because a fixed dependent variable is assigned with a fixed value of an independent variable. For example, damages resulting from a given water level in a given structure are treated as deterministic. In this case, however, an estimation error is applied to the result to reflect imprecision in the physical survey of properties in the floodplain.

Finally, other relationships relate to empirical correlations. These relationships might be called *statistical* because they are treated as statisti-

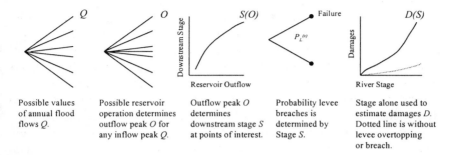

Possible values of annual flood flows Q.	Possible reservoir operation determines outflow peak O for any inflow peak Q.	Outflow peak O determines downstream stage S at points of interest.	Probability levee breaches is determined by Stage S.	Stage alone used to estimate damages D. Dotted line is without levee overtopping or breach.

FIGURE 4.1. Event tree describing the transformation of flood discharge into damage. Some of the steps in this process are deterministic; others are stochastic. SOURCE: NRC (1995).

cal regression functions. For example, river stage as a function of flood-flow is treated as statistical. The estimation error applied to the damage–stage relationship mentioned previously is also modeled as a statistical error.

For each phase of Figure 4.1 there is thus a mathematical relation-ship that translates input variables to output variables, and each relation-ship introduces uncertainties. Some of the uncertainties derive from natural variations, others from engineering calculations, and still others from statistical estimation. For each relationship in the event tree of Fig-ure 4.1 there is a set of parameters that define the corresponding equation or curve. The values of these parameters are uncertain. Consider two phases in the event tree (which shows five phases) as illustrative; the flood–frequency relationship on the left (the first phase), and the stage–discharge relationship in the middle (the third phase).

The possible values of the annual flood flow Q are represented by an exceedance probability distribution. These flood flows are assumed to be naturally variable and are describable with probabilities. This re-quires a set of parameters to specify the distribution shape and location along the axis of river flows (discharges). The parameters used are usu-ally the mean, variance, and skewness coefficient of the logarithms of the flows. On the other hand, the shape and the location of the probability distribution are themselves uncertain because of imperfect knowledge about which distribution model to fit to historical data and about the best values of distribution parameters for that model. Thus, the flood flows involve both natural variability and knowledge uncertainty.

The stage–discharge relationship is represented by a regression equation. This requires a set of parameters to describe the shape and location of the relationship on a graph of stage vs. discharge. The parameters used are usually an intercept, slope, and maybe some form of shape factor. To some extent, the stage–discharge relationship reflects natural variability over time or within the river reach (e.g., variability caused by water temperature, by scour and deposition or by stages of tributaries). On the other hand, the regression equation is estimated from limited data. The shape and the location of the regression curve are themselves uncertain because of imperfect knowledge about which equation to fit to historical data and about the best values of regression parameters for that equation. As with the flood–frequency curve, there is imperfect knowledge about which probability distribution model to fit to natural variations of historical stage data, about the regression curve, and about appropriate values of the distribution parameters for that model. Thus, the stage–discharge relationship also involves both natural variability and knowledge uncertainties. The largest knowledge uncertainties are for uncommon, extremely large floods.

The Corps's objective in flood damage reduction studies is to determine the expected annual damage (EAD) along a section of river caused by possible floods, and to compare changes in those damages as a function of project alternatives. The Corps's method for such calculations starts with flood discharge, Q, which is equaled or exceeded—on average—once in T years. T is said to be the return period of the flood discharge Q. Corresponding to this return period T is a probability p that the discharge Q is equaled or exceeded in any given year. This annual exceedance probability is the reciprocal of the return period, T, and is given by

$$p = \frac{1}{T}.$$
(4.1)

For a flood of annual probability p, a corresponding value of flood damage $D(p)$ can be estimated. This is based on the depth of inundation of the floodplain and on the value of the inundated structures. The EAD is the average value of such damages taken over floods of all different annual exceedance probabilities and over a long period of years. Stated mathematically, the EAD is,

$$EAD = \int_0^1 D(p)dp\,. \qquad\qquad (4.2)$$

The current Corps method divides the calculation of Equation 4.2 into three steps (Figure 4.2):

1. *determining flood frequencies*, which describe the probability of floods equal to or greater than some discharge Q (i.e., volume of flow) occurring within a given period of time—shown in the upper-right panel of Figure 4.2,

2. *determining stage–discharge relations*, which describe how high the flow of water in a reach of river (the stage) might be for a given volume of flow (discharge)—shown in the upper-left panel of Figure 4.2, and

3. *determining damage–stage relations*, which describe the amount of damage that might occur, given a certain height of flow—shown in the lower left panel of Figure 4.2.

The lower-right panel of Figure 4.2 relates annual damage to exceedance probability, p. The shaded area under the curve is the expected annual damage, given by Equation 4.2. To find the damage for a given probability p, the discharge Q_T for that probability is first taken from the flood–frequency curve, given in the upper-right panel. Then the stage height (water surface elevation) for that discharge, H, is found from the upper-left panel. From the value of H, the damage D for stage height is found from the lower-left panel. By plotting this damage on the lower-right panel for the given probability, and by repeating this process for a sequence of flood probabilities, the damage–frequency curve is established. This curve is then integrated to give expected annual damage. In the Corps's method, annual exceedance probabilities of $p = 0.5, 0.2, 0.1, 0.04, 0.02, 0.01, 0.004,$ and 0.002 are the values used in the computation.

Uncertainties enter the calculations in each step of the analysis and are propagated from one step to the next, ultimately accumulating in the EAD—the estimate of damages that might occur in a given year. These uncertainties in damage estimates are expressed as a frequency curve of damages—analogous to a frequency curve of flooding—describing the probability of damages of a given magnitude being exceeded in a given period of time (e.g., annually). This frequency curve of damages is

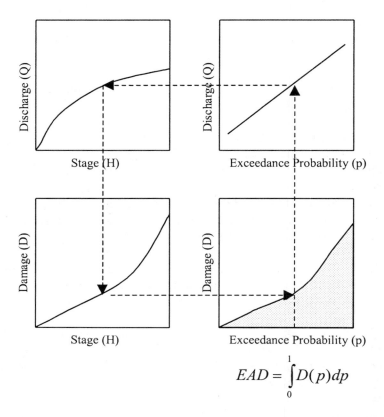

$$EAD = \int_{0}^{1} D(p)dp$$

FIGURE 4.2   Basis of the Corps's computation of expected annual damage (EAD). The logic of this figure flows counterclockwise starting from the upper right panel and ending in the lower right panel.
SOURCE: Adapted from Moser (1997).

shown in the lower right panel of Figure 4.2. The Corps's conceptual approach to modeling flood hazard and associated damages—using the relationships between flood frequency, stage–discharge, and damage–stage—is consistent with longstanding scientific understanding.

The computational procedure in the Corps's method uses Monte Carlo sampling to perform numerical integration of the damage exceedance probability curve for a damage reach. The damage-exceedance probability function is obtained from the discharge-exceedance probability, stage-discharge, and damage-stage functions. The numerical integration is necessary because the damage-exceedance probability function in not defined by a continuous analytic function. In this procedure, (pseudo-) random numbers are used to generate a single realization of each of the three relationships: discharge-exceedance probability function, stage-discharge function, and stage-damage function. From these, a single realization of the damage-exceedance probability curve is calculated. The expected annual damage (EAD) is calculated for this realization by integrating the damage-exceedance probability curve. This process is repeated many times and statistically averaged. The numerical results can be made arbitrarily precise, at least from a statistical point of view, by increasing the number of realizations calculated in this way.

The Corps's approach is a reasonable risk analysis procedure that deserves consideration for wider adoption in the flood management community. It provides a mechanism for combining uncertainty in estimating flood discharge and stage with the inherent risk of different flood severities, to give overall risk measures of the system's engineering performance that are more complete than those customarily used.

RISK ANALYSIS

Determination of EAD in Equation 4.2, as historically performed, considered the range of flood magnitudes that could occur, but it did not consider uncertainties in hydrologic, hydraulic, or economic information used in the damage calculation. The curves of Figure 4.2 were treated as known. The traditional approach, as illustrated by Figure 4.2, does not tell us how sure we can be that the calculated expected annual damages will not be exceeded. This is because there are uncertainties in the probability distribution of annual peak flood flows, in the relationship between flood flow and flood stage, and in the relationship between flood stage and economic damage. Just how accurate is the calculated estimate of EAD damage? The Corps's new risk analysis for flood damage assessment attempts to quantify both the natural variability and the knowledge uncertainty in the above procedure.

Risk analysis provides a means of estimating a range of expected annual flood damages, each of which is associated with a level of assurance that it will not be exceeded. Similarly, risk analysis can be employed to calculate the range of expected probabilities of levee failure, each of which is associated with a level of assurance that that probability will not be exceeded. Consider the upper-right quadrant in Figure 4.2. The peak flow expected once in 100 years, on average, is the flow corresponding to an annual exceedance probability of $p = 0.01$. How sure can we be that if we protect ourselves from a flow of that particular magnitude, we will actually be protecting ourselves from all flows that occur less frequently, on average, than once in 100 years? Risk analysis addresses such questions.

Assume that the probability distribution capturing the uncertainty about the probability of exceedance of the peak flows at the potential damage site (as shown in the upper-right quadrant of Figure 4.2) was determined as part of a risk analysis. Figure 4.3 shows a portion of that function. To be, say, 90 percent sure that protection is provided for the 100-year return period flow, Q_{100} depends on the uncertainty in the estimated flow probabilities (see Figure 4.3), stage–discharge relationship, and in the levee system reliability.

Risk, uncertainty, and variability are inherent in flood damage reduction planning. There is uncertainty in any forecast of stream or river flood flows and the resulting damage simply because we cannot know enough about all the factors that contribute to them. Uncertainties in the stage–damage function in the lower-left quadrant of Figure 4.2 include:

• economic activities and the economic condition or value of the property on the floodplain during a flood,
• warning time and response of floodplain inhabitants,
• velocity of the floodwaters and the amount of mud and debris, and
• time required to repair damaged property.

Uncertainties in the discharge–stage function in the upper-left quadrant of Figure 4.2 include:

• physical characteristics of channel,
• winds that may affect flood stages associated with given flows, and
• vegetation, debris, and other obstructions including ice in the channel.

Uncertainties in the probability of exceedance distribution of annual peak flows in the upper-right quadrant of Figure 4.2 include:

• limited data from which to statistically estimate hydrographs,
• when and how severe a rain storm or other event (e.g., upstream dam failure or upstream or downstream levee failure) may be that could

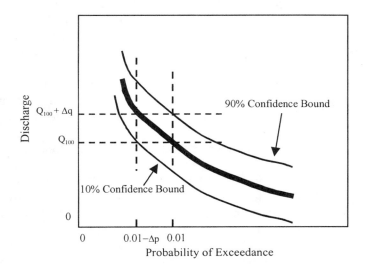

FIGURE 4.3. The darkest of these three curves represents the expected probability of exceedance of peak flow derived from a flood-frequency analysis. The two lighter lines that bound this curve define the 10% and 90% confidence bounds, respectively, around the expected probability of exceedance. Q_{100} represents discharge that has a 0.01 probability of being exceeded, while $Q_{100}+\Delta q$ is the discharge level for which there is a 90 percent confidence that the 100-year flow will not exceed this level. In terms of expected probability, this discharge has a $0.01-\Delta p$ probability of being exceeded. This approach corresponds to a safety factor that accounts for the uncertainty in the hydrologic analysis and the assessment of flood frequencies.

result in a flood,

• duration and distribution of rainfall on an area draining a potential flood damage site, and the precise rainfall–runoff and flow routing events (such as watershed topography, land use and cover, soil moisture content) that exist during such a storm,

• the likelihood of levee or upstream dam (structural) failures or operator (nonstructural) "failures" at upstream dams, and

• actions (temporary measures) taken upstream during a flood to protect upstream sites.

Peak flow records are commonly used to estimate the chance of a flood of a given or greater magnitude. But such estimates are uncertain for at least two reasons. One is the limited (and perhaps inaccurate) number of observations of past peak flows used to estimate the likelihood of equaling or exceeding a particular peak flow in any year. The other is the changing and varying character of the drainage basin that influences the peak flow resulting from a specific rainfall. In many cases the probability distribution of peak flows is changing, even assuming, perhaps incorrectly at least in the long run, that the probabilistic character of the rainfall is not.

The Corps's search for a better method of quantifying flood risk was prompted by issues related to riverine levee freeboard, described at a workshop held in Monticello, Minnesota (USACE, 1991a). There were concerns that arbitrarily defined safety margins were not explicitly related to the causes of uncertainty in levee performance and that the additional height (freeboard) required to meet these safety margins was not properly accounted for in the evaluation of project benefits. At the Monticello meeting a methodology was presented for quantifying uncertainty in discharge, stage, and damage, (Davis, 1991) from which the current procedure has evolved. At first, the risk analyses for flood damage assessment were computed with a spreadsheet with an add-in, commercial-off-the-shelf program, as described by Davis (1991). In January 1998 the Corps's Hydrologic Engineering Center (HEC) released Version 1.0 of the HEC-Flood Damage Assessment computer program (HEC-FDA) (HEC, 1998a), which provided an improved Windows program for carrying out the computations.

MONTE CARLO SIMULATION

Figure 4.4 shows a sequence of three graphs describing uncertainties in discharge, stage, and damage. The uncertainty in these quantities is signified by the probability distributions and dashed confidence-limit lines drawn around each curve and also by the dash–dot lines on each graph, which are possible alternative locations of the curves. The first graph shows the uncertainty in discharge Q for a given exceedance probability p, $f_I(Q|p)$. This uncertainty, $f_I(Q|p)$, can be specified rather precisely using the noncentral t distribution if the flood–frequency curve is described by the log normal distribution, and approximately so if the flood–frequency curve is described by other methods. A flood–frequency curve based on the log normal distribution is completely speci-

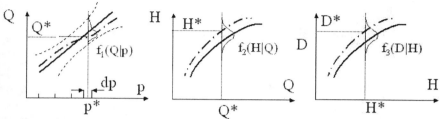

FIGURE 4.4 Uncertainties in the discharge (*Q*), stage (*H*), and damage (*D*) relationships that are part of the risk analysis method. The *first figure* shows the relationship between discharge vs. exceedance probability (*p*). The solid curve is the best estimate of this relationship, while the dash-dot curve represents one potential realization that the actual curve might assume. The dotted lines show probability contours of the function f1 (Q|p), which is the probability density function describing uncertainty in discharge for a given exceedance probability. The *second figure* shows the relationship between discharge and water height (i.e., the rating curve). The solid line is the best estimate and the dash-dot curve is one potential realization that the actual curve might assume. The function f2(H|Q) is the probability density function of height given discharge. The *third figure* shows the relationship between damage and water height. The solid line is the best estimate of this relationship and the dash-dot curve is one potential realization that the actual curve might assume. The function f3(D|H) is the probability density function of damage given water height.

fied by its mean and standard deviation, so if the mean and standard deviation are varied, then different flood–frequency curves result. The statistical uncertainty in the mean and standard deviation can be quantified based on the number of data values used in calculating them. By using Monte Carlo simulation to generate different means and standard deviations and then plotting the resulting flood–frequency curves, different "realizations" of the flood–frequency curve are defined, one example of which is shown by the dash–dot line in $f_1(Q|p)$ in Figure 4.4. It should be noted that several methods could be used for risk value computations, and that the Corps uses the Monte Carlo method.

Similarly, the uncertainty in the rating curve, or relationship between the stage height H and the discharge Q, is symbolized by the probability distribution around the $f_2(H|Q)$ curve in Figure 4.4 and by the alternative realization of that curve shown by the dash–dot line in that figure. In this case, the Monte Carlo analysis uses only a single random variable that serves to vertically displace the average rating curve.

Finally, the uncertainty in the damage–stage function is symbolized by the probability distribution around the $f_3(D|H)$ curve in Figure 4.4 and the alternative realization of that curve, given by the dash–dot line. This function combines the effects of several different kinds of uncertainty, including the likelihood of levee failure, uncertainty in elevations of structures in the floodplain, lack of knowledge of the degree of flood damage for a given depth of inundation in a structure, and uncertainty in property and content values within the structures.

Monte Carlo simulation is used to generate new realizations of each of the three curves in Figure 4.4. For each of these sets of realizations, a new value of expected annual damages (EAD) is found by the same process as described in Figure 4.2, as if each realization were the true value of the curve. In other words, for each exceedance probability interval dp, a representative exceedance probability $p*$ is used, from which the flood discharge $Q*$ is found using the flood–frequency curve, the corresponding stage height $H*$ from the discharge–stage curve, and the consequential damage $D*$ from the damage–stage function. By continuing this process across the exceedance probability axis and then integrating the results using Equation 4.2, the EAD is found. The Monte Carlo simulation is continued for a few thousand cycles of generating realizations and computing EAD, until the statistics of the EAD values are sufficiently accurate. This form of Monte Carlo simulation is more sophisticated than the simpler approach of simply generating a flood, finding the stage and damage, generating a new flood, finding the stage and damage, and so on. Thus, the Monte Carlo simulation uses random numbers to perturb the relationships linking the key variables, rather than to generate random floods and examine their consequences.

As the Corps's risk analysis methods evolve, it is possible that a direct Monte Carlo approach will become impractical. The approach requires large numbers of repetitive calculations, and should the analysis models become more involved, alternative calculation approaches may be desirable.

ASSESSMENT OF ENGINEERING PERFORMANCE

The engineering performance of a flood damage reduction project is measured by the probability that the land to be protected by the project will be flooded in any given year. Such probabilities are estimated for each damage reach in the project and consider hydrologic, hydraulic, and geotechnical uncertainties. Engineering performance is not concerned with estimates of economic damage, which are assessed separately. The performance measures require the definition of a target stage for each damage reach, as shown in Figure 4.5. The target stage is defined as the water surface elevation in a reach at which significant economic damage occurs. To determine the target stage, the damage–frequency curve is obtained for the damage reach by the process shown in Figure 4.2, transforming the flood–frequency curve through the damage–stage curve without consideration of errors in these curves. The 1 percent chance of flooding damage is found from this damage–frequency curve. A fraction of this damage (usually 10 percent) is taken and is used to determine the corresponding stage from the damage–stage curve, which then becomes the target stage for the reach.

As Figure 4.6 shows, engineering performance can be stated in two ways—either as a risk of failure, measured by the exceedance probability of a target stage, or as a reliability, measured by the nonexceedance probability of the target stage. Performance can also be measured by conditional probabilities dependent on the occurrence of a flood of a given severity (e.g., the 100-year flood) or dependent on the annual probabilities integrated over all the floods that could occur within a given year. In the Corps's method, the two main engineering performance measures combine the two sets of distinctions into the following measures:

- *annual exceedance probability*—the probability that the target stage will be exceeded in any year considering all potential floods and
- *conditional nonexceedance probability*—the probability that the target stage will not be exceeded given a specific flood severity.

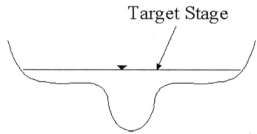

Target Stage

FIGURE 4.5 Definition of a target stage used in assessing engineering performance.

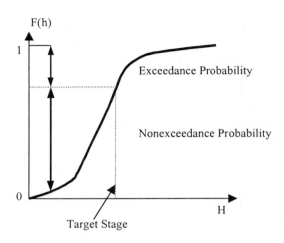

FIGURE 4.6 Exceedance and nonexceedance probabilities.

The assessment of engineering performance to create the annual exceedance and conditional nonexceedance probabilities is carried out as illustrated in Figure 4.7. The first two panels in Figure 4.7 show the flood–frequency curve and the stage–discharge curve. In each cycle of the Monte Carlo procedure, a new realization of the flood–frequency curve $f_1(Q|p)$ and the stage- damage curve $f_2(H|Q)$) is generated, where H represents stage height and Q represents discharge. The flood–frequency curve is defined at discrete intervals of annual flood probability, p (p = 0.5, 0.2, 0.1, 0.04, 0.02, 0.01, 0.004, 0.002). If a particular value of p is chosen, say p*, the corresponding flood discharge Q^* can be found from the flood–frequency curve, and the resulting stage height H^* can be found from the stage–discharge curve. By combining these pairs of (H^*, p^*) values, a stage–frequency curve can be constructed, $f_3(H|p)$, as shown in the third panel in Figure 4.7.

The annual exceedance probability, p_e, is estimated from the stage–frequency curve as that probability corresponding to the target stage for the damage reach. This computation is repeated for N cycles of Monte Carlo simulations, and the expected value of the annual exceedance probability is found as the average over the N sample values from the

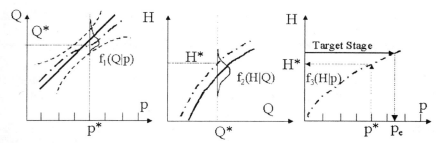

FIGURE 4.7 Computation of risk measures using a target stage.

simulations. The median annual exceedance probability is the 50 percent value of this distribution.

Conditional nonexceedance probabilities are also determined from the stage–frequency curve. For each value of $p*$, there corresponds an $H*$, determined in the manner just described. After all the Monte Carlo simulations are complete, a set of N values of $H*$ exists, of which a subset, n, have stages not exceeding the target stage. The conditional nonexceedance probability of the target stage is given by n/N. The HEC-FDA program presents such conditional nonexceedance probabilities for each damage reach for annual event probabilities of 0.1, 0.04, 0.02, 0.01, 0.004, and 0.002, shown in the following chapter in Figure 5.10. Figure 5.10 also shows the chance that the target stage will be exceeded at least once in 10, 25, and 50 years, computed as $1 - (1-p_e)^n$, where $n = 10, 25,$ or 50, respectively.

GEOTECHNICAL RELIABILITY

Even if not overtopped by floods, levees may fail for geotechnical reasons. The Corps's risk analysis procedure incorporates the chance of such failures through a geotechnical reliability model. This model leads to a relationship between water height and probability of geotechnical failure, which is then applied individually to each damage reach of river. The logic of this calculation is that damages accrue in one of two ways— either the river becomes high enough that a levee is overtopped, or even though the river does not overtop a levee, it is high enough to cause geotechnical failure.

The Corps's geotechnical reliability model is a sound first step in balancing scientific understanding with the practical needs of planning studies and risk analysis. This is a difficult problem. The geotechnical performance of a levee depends on local soil conditions and construction details, neither of which are known in detail during the planning study. Many of the levees of concern to risk analysis studies were not designed by the Corps and are neither owned nor maintained by federal agencies.

The Corps's original geotechnical reliability model is a simple relationship based on two critical stage heights for the levee: the probable failure point (PFP) and the probable nonfailure point (PNP) (USACE, 1991b). The probable failure point is the stage height associated with a high probability of failure. Numerically, this probability is set at 0.85. The probable nonfailure point is the stage height associated with a negligible probability of failure. Numerically, this probability is set at 0.15. A line is drawn between the PFP and PNP, as shown in Figure 4.8. These points are assessed for local conditions of the project area, and they may change from reach to reach. To avoid complications arising from failures in multiple locations on a long levee, the Corps models the damage such that the reach covers the whole length of the levee under consideration. Thus, the risk function in Figure 4.8 refers to the chance of failure at only the weakest point over this reach. This original model appears to be still widely used in the Corps's district offices.

The original reliability model has been updated (USACE, 1999b) to reflect more sophisticated understanding of geotechnical performance. The updated model considers multiple modes of geotechnical failure, including underseepage, through seepage, and strength instability. This results in a composite curve that varies smoothly between probabilities of 0 and 1, rather than being anchored to a probable failure point and a probable nonfailure point (Figure 4.9).

Although the updated model is based on better scientific understanding of levee performance, the numerical difference in risk analysis results compared to the initial model may not be large. The updated model, however, supports a more complete geotechnical analysis and should to replace the initial model.

The risk measures for engineering performance, including geotechnical reliability, are calculated as shown in Figure 4.10. The first panel of this figure shows the stage–frequency curve, $f_3(H|p)$, from Figure 4.7, which is determined from the hydrologic and hydraulic models and their attendant uncertainties. For each frequency value ($p^* = 0.5, 0.2, 0.1, 0.04, 0.02, 0.01, 0.004, 0.002$), a corresponding stage height H^* is de-

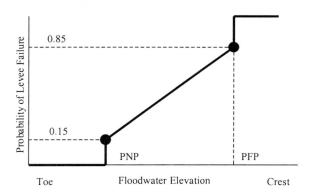

FIGURE 4.8   Two-point model of geotechnical levee reliability. PFP is
probable failure point; PNP is the probable nonfailure point.

termined. The middle panel of Figure 4.10 is the risk of levee failure R
as a function of stage, $f_4(R|H)$, derived from Figure 4.8 or 4.9 using this
function. For the given value of H^*, the corresponding risk of failure R^*
is determined. The pairs of values (R^*, p^*) are combined to form a risk–
frequency curve, $R(p)$, as shown in the last panel of Figure 4.10. The
annual exceedance probability p_e including geotechnical uncertainty is
then found in an analogous manner to the expected annual damage using
Equation (4.3):

$$p_e = \int_0^1 R(p)dp . \qquad (4.3)$$

The conditional nonexceedance probability for any given value of p^* is
simply $1 - R^*$ in Figure 4.10.

   By repeating this computation many times using a Monte Carlo
simulation, a set of of p_e and $1 - R^*$ values is obtained. The p_e values are
averaged to find the expected value of the annual exceedance probability,
and the $1 - R^*$ values are averaged to find the expected value of the con-
ditional-nonexceedance probability.   Consider, for example, the construction of a levee with a specified conditional nonexceedance probability of, for example, 90 percent or 95 percent for a 100-year flood. In

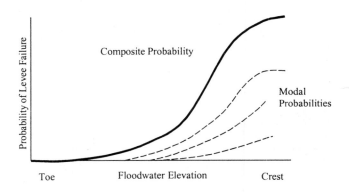

FIGURE 4.9 Continuous model of geotechnical levee reliability.

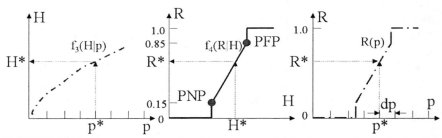

FIGURE 4.10 Computation of risk measures including geotechnical reliability.

this instance, the above procedure is executed with a specified value of $p^* = 0.01$, and the levee height is raised or lowered until the required conditional nonexceedance probability is obtained.

The Corps's analysis combines uncertainty about the parameters and analysis models with the variability inherent in natural systems. Applications are discussed in Chapter 5.

5

Case Studies

This chapter illustrates the Corps of Engineers's application of risk analysis by reviewing two Corps flood damage reduction projects: Beargrass Creek in Louisville, Kentucky, and the Red River of the North in East Grand Forks, Minnesota, and Grand Forks, North Dakota. The Beargrass Creek case study describes the entire procedure of risk-based engineering and economic analysis applied to a typical Corps flood damage reduction project. The Red River of the North case study focuses on the reliability of the levee system in Grand Forks, which suffered a devastating failure in April 1997 that resulted in more than $1 billion in flood damages and related emergency services.

The Corps of Engineers has used risk analysis methods in several flood damage reduction studies across the nation, any of which could have been chosen for detailed investigation. Given the limits of the committee's time and resources, the committee chose to focus upon the Beargrass Creek and Red River case studies for the following reasons: committee member proximity to Corps offices, a high level of interest in these two studies, and the availability of documentation from the Corps that adequately described their risk analysis applications.

Differences in approaches taken at Beargrass Creek and along the Red River of the North to reducing flood damages are reflected in these studies. At Beargrass Creek, the primary flood damage reduction measures were detention basins; at the Red River of the North, the primary measures were levees. The Corps uses rainfall-runoff models in nearly all of its flood damage reduction studies to simulate streamflows needed for flood-frequency analysis, and a rainfall-runoff model was employed in the Beargrass Creek study. In the Red River study, however, the goal was to design a system that would, with a reasonable degree of reliability, contain a flood of the magnitude of 1997's devastating flood. The Corps focused on traditional flood-frequency analysis and manipulated the frequency curve at a gage location to derive frequency curves at other locations (vs. using a rainfall-runoff model to derive those curves).

BEARGRASS CREEK

In 1997 the Corps held a workshop (USACE, 1997b) at which experience accumulated since 1991 in risk analysis for flood damage reduction studies was reviewed. O'Leary (1997) described how the new procedures had been applied in the Corps's Louisville, Kentucky, district office. In particular, O'Leary described an application to a flood damage reduction project for Beargrass Creek, economic analyses for which were done both under the old procedures without risk and uncertainty analysis and under the new procedures that include those factors. Conclusions of the Beargrass Creek study are summarized in two volumes of project reports (USACE, 1997c,d). These documents, plus a site visit to the Louisville district by a member of this committee, form the basis of this discussion of the Beargrass Creek study. The Beargrass Creek data are distributed with the Corps's Hydrologic Engineering Center Flood Damage Assessment (HEC-FDA) computer program for risk analysis as an example data set. The Beargrass Creek study is also used for illustration in the HEC-FDA program manual and in the Corps's Risk Training course manual. Although there are variations from study to study in the application of risk analysis, Beargrass Creek is a reasonably representative case with which to examine the methodology.

As shown Figure 5.1, Beargrass Creek flows through the city of Louisville, Kentucky, and into the Ohio River on its south bank. The Beargrass Creek basin has a drainage area of 61 square miles, which encompasses about half of Louisville. The basin currently (year 2000) has a population of about 200,000. This flood damage reduction study's focal point is the lower portion of the basin shown in Figure 5.1—the South Fork of Beargrass Creek and Buechel Branch, a tributary of the South Fork.

Locally intense rainstorms (rather than regional storms) cause flooding in Beargrass Creek. A 2-year return period storm causes the creek to overflow its banks and produces some flood damage. Under existing conditions, the Corps estimates that a 10-year flood will impact

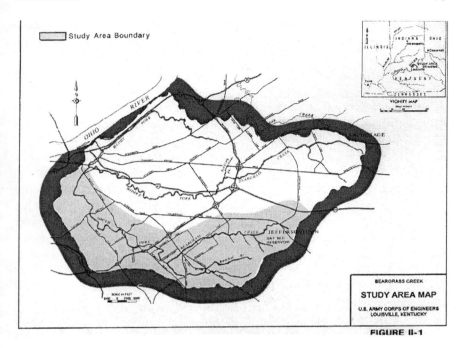

FIGURE 5.1 The Beargrass Creek basin in Louisville, Kentucky. SOURCE: USACE (1997a) (Figure II-1).

about 300 buildings and cause about $7 million in flood damages, while a 100-year flood will impact about 750 buildings and cause about $45 million in flood damages (USACE, 1997c). The expected annual flood damage under existing conditions is approximately $3 million per year.

Flood Damage Reduction Measures

Beargrass Creek has several flood damage reduction structures, the most notable of which is a very large levee at its outlet on the Ohio River (Figure 5.2a). This levee was built following a disastrous flood on the Ohio in January 1937, and the levee crest is an elevation of 3 feet above the 1937 flood level on the Ohio River. During the 1937 flood it was reported that "at the Public Library, the flood waters reached a height such that a Statue of Lincoln appeared to be walking on water!" (USACE, 1997b, p. III-2). Near the mouth of Beargrass Creek, a set of gates can be closed to prevent water from the Ohio River from flowing back up into Louisville. In the event of such a flood, a massive pump station with a capacity of 7,800 cubic feet per second (cfs) is activated to discharge the flow of Beargrass Creek over the levee and into the Ohio River.

Between 1906 and 1943, a traditional channel improvement project was constructed on the lower reaches of the South Fork of Beargrass Creek. It consists of a concrete lined rectangular channel with vertical sides, with a small low-flow channel down the center (Figure 5.2b). The channel's flood conveyance capacity is perhaps twice that of the natural channel it replaced, but the concrete channel is a distinctive type of landscape feature that environmental concerns will no longer permit. Other structures have been added since then, including a dry bed reservoir completed in 1980, which functions as an in-stream detention basin during floods.

The proposed flood damage reduction measures for Beargrass Creek form an interesting contrast to traditional approaches. The emphasis of the proposed measures is on altering the natural channel as little as possible and detaining the floodwaters with detention basins. These basins are either located on the creek itself or more often in flood pool areas adjacent to the creek into which excessive waters can drain, be held for a few hours until the main flood has passed, and then gradually return to the creek. Figure 5.2c shows a grassed detention pond area with a concrete weir (in the center of the picture) adjacent to the creek. Figure 5.2d shows Beargrass Creek at this location (a discharge pipe from the pond is visible on the right side of the photograph). Water flows from the creek into the pond over the weir and discharges back into the creek through the pipe. The National Economic Development flood damage reduction alternative on Beargrass Creek called for a total of eight detention basins, one flood wall or levee, and one section of modified channel. Other alternatives such as flood-proofing, flood warning systems, and enlargement of bridge openings were considered but were not included in the final plan.

The evolution of flood damage reduction on Beargrass Creek represents an interesting mixture of the old and the new—massive levees and control structures on the Ohio River, traditional approaches (the concrete-lined channel) in the lower part of the basin, more modern in-stream and off-channel detention basins in the upstream areas, and local

(a) Levee on the Ohio River

(b) Concrete-lined channel

(c) Detention pond

(d) Beargrass Creek at the detention pond

FIGURE 5.2   Images of Beargrass Creek at various locations: (a) the levee on the Ohio River, (b) a concrete-lined channel, (c) a detention pond, and (d) the Beargrass Creek at the detention pond.

channel modifications and floodwalls. Maintenance and improvement of stormwater drainage facilities in Beargrass Creek are the responsibility of the Jefferson County Metropolitan Sewer District, which is the principal local partner working with the Corps to plan and develop flood damage reduction measures.

In some locations, development has been prohibited in the floodway; but in other places, buildings are located adjacent to the creek. The Corps's feasibility report includes the following comments: "Urbanization continues to alter the character of the watershed as open land is converted to residential, commercial and industrial uses. The quest for open area residential settings in the late 1960s and early 1970s caused a tremendous increase in urbanization of the entire basin. Several developers have utilized the aesthetic beauty of the streambanks as sites for residential as well as commercial developments. This has resulted in increased runoff throughout the drainage area as development has occasionally encroached on the floodplain and, less frequently, the floodway" (USACE, 1997b, p. II-2).

Damage Reaches

To conduct the flood damage assessment, the two main creeks— South Fork of Beargrass Creek and Buechel Branch—are divided into damage reaches. Flood damage and risk assessment results are summarized for each damage reach, and the expected annual damage for the project as a whole is found by summing the expected annual damages for each reach. As shown in Figure 5.3, the South Fork was divided into 15 damage reaches and the Buechel Branch into 5 reaches (a sixth damage reach on Buechel Branch is not shown in this figure). Approximately 12 miles of Beargrass Creek, and 2.2 miles of Buechel Branch are covered by the these damage reaches. The average length of a damage reach is thus 0.8 miles for the South Fork of the Beargrass Creek, and the average length for Buechel Branch is 0.4 miles. The shorter reaches on Buechel Branch are adjacent to similarly short, upstream reaches in Beargrass Creek where most flood damage occurs. Longer damage reaches are used downstream on Beargrass Creek where less damage occurs.

The highest expected annual flood damage is on Reach SF-9 on the upper portion of the South Fork of Beargrass Creek. Results from this damage reach are used for illustrative purposes at various points in this chapter.

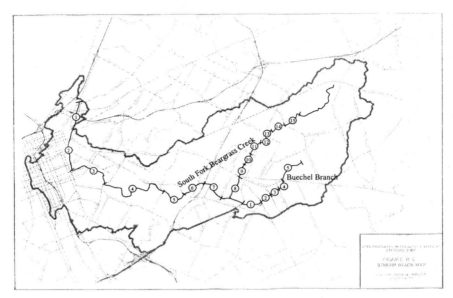

FIGURE 5.3   Damage reaches on the South Fork of Beargrass Creek and Buechel Branch. SOURCE: USACE (1997a) (Figure III-3).

Flood Hydrology

Most of the flood damage reduction measures being considered are detention basins, which diminish flood discharge by temporarily storing floodwater. It follows that the study's flood hydrology component has to be conducted using a time-varying rainfall–runoff model because this allows for the routing of storage water through detention basins. In this case, the HEC-1 rainfall–runoff model from the Corps's Hydrologic Engineering Center (HEC) was used to quantify the flood discharges. The Hydrologic Engineering Center has subsequently released a successor rainfall-runoff model to HEC-1, called HEC-HMS (Hydrologic Modeling System), which can also be used for this type of study (HEC, 1998b).

In each damage reach, and for each alternative plan considered, the risk analysis procedure for flood damage assessment requires a flood–frequency curve defining the annual maximum flood discharge at that location which is equaled or exceeded in any given year with a given probability. In this study all these flood–frequency curves were produced through rainfall–runoff modeling. In other words, a storm of a given return period was used as input to the HEC-1 model, the water was routed through the basin, and the magnitude of the discharge at the top end of each damage reach was determined (Corps hydrologists have assumed, based on experience in the basin, that storms of given return periods produce floods of the equivalent return period). By repeating this exercise for each of the annual storm frequencies to be considered, a flood–frequency curve was produced for each damage reach. There are eight standard annual exceedance probabilities normally used to define this frequency curve: p = 0.5, 0.2, 0.1, 0.04, 0.02, 0.01, 0.004, and 0.002, corresponding to return periods of 2, 5, 10, 25, 50, 100, 250, and 500 years, respectively. In this study, because even small floods cause damage, a 1-year return period event was included in the analysis and assigned an exceedance probability of 0.999.

Considering that there are 21 damage reaches in the study area and 8 annual frequencies to be considered, each alternative plan considered requires the development of 21 flood–frequency curves involving 168 discharge estimates. During project planning, as dozens of alternative components and plans were considered, the sheer magnitude of the tasks of hydrologic simulation and data assembly becomes apparent.

The hydrologic analysis is further complicated by the fact that the design of detention basins is not simply a cut-and-dried matter. A basin designed to capture a 100-year flood requires a high–capacity outlet structure. Such a basin will have little impact on smaller floods because the outlet structure is so large that smaller events pass through almost unimpeded. If smaller floods are to be captured, a more confined outlet structure is needed, which in turn increases the required storage volume for larger floods. This situation was resolved in the Beargrass Creek study by settling on a 10-year flood as the nominal design event for sizing flood ponds and outlet works. The structures designed in this manner were then subjected to the whole range of floods required for the economic analysis.

Rainfall–Runoff Model

The HEC-1 model was validated by using historical rainfall and runoff data for four floods (March 1964, April 1970, July 1973, February 1990). Modeling results were within 5 percent to 10 percent of observed flows at two U.S. Geological Survey (USGS) streamflow gaging stations: South Fork of Beargrass Creek at Trevallian Way and Middle Fork of Beargrass Creek at Old Cannons Lane, which have flow records beginning in 1940 and 1944, respectively, and continuing to the present. A total of 42 subbasins were used in the HEC-1 model, and runoff was computed using the U.S. Soil Conservation Service (renamed the Natural Resources Conservation Service in 1994) curve number loss rates and unit hydrographs. The Soil Conservation Service curve numbers were adjusted to allow the matching of observed and modeled flows for the historical events. A 6-hour design storm was used, which is about twice the time of concentration of the basin. The design storm duration chosen is longer than the time of concentration of the basin so that the flood hydrograph has time to rise and reach its peak outflow at the basin outlet while the storm is still continuing. If the design storm is shorter than the time of concentration, rainfall could have ceased in part of the basin before the outflow peaks at the basin outlet. The storm rainfall hydrograph was based on National Weather Service 1961 Technical Paper 40 (NWS, 1961) and on a Soil Conservation Service storm hydrograph, and a 5-minute time interval of computation was used for determining the design discharges.

There is a long flood record of 56 years of data (1940–1996) available in the study area (USGS gage on the South Fork of Beargrass Creek at Trevallian Way). A comparison was made of observed flood frequencies at this site with those simulated by HEC-1, with some adjustment of the older flood data to allow for later development. Traditional flood frequency analysis of observed flow data had little impact in the study. This may have been the case because there was only one gage available within the study area, or because the basin has changed so much over time that the flood record there does not represent homogeneous conditions. Furthermore, the alternatives mostly involve flood storage, which requires computation of the entire flood hydrograph, not just the peak discharge.

Uncertainty in Flood Discharge

Uncertainty in flood hydrology is represented by a range in the estimated flood–frequency curve at each damage reach. In the HEC-FDA program, there are two options for specifying this uncertainty: an analytical method based on the log-Pearson distribution and a more approximate graphical method. The log-Pearson distribution is a mathematical function used for flood–frequency analysis, the parameters of which are determined from the mean, standard deviation, and coefficient of skewness of the logarithms of the annual maximum discharge data. The graphical method is a flood frequency analysis performed directly on the annual maximum discharge data without fitting them with a mathematical function. In this case the graphical method was used with an equivalent record length of 56 years of data, the length of the flood record of the USGS gage station at Trevallian Way at the time of the study. Figure 5.4 shows the flood–frequency curve for damage reach SF-9 on the South Fork of Beargrass Creek, with corresponding confidence limits based on ± 2 standard deviations about the mean curve.

The confidence limits in this graph are symmetric about the mean when the logarithm to base 10 of the discharge is taken, rather than the discharge itself. This can be expressed mathematically as:

$$\log Q = \overline{\log Q} + K\sigma_{\log Q}, \qquad (5.1)$$

where Q is the discharge value at the confidence limit, $\overline{\log Q}$ is the expected flood discharge, $\sigma_{\log Q}$ is the standard deviation (shown in the rightmost column of Table 5.1), and K is the number of standard deviations above or below the mean that the confidence limit lies. Because these confidence limits are defined in the log space, it follows that they are not symmetric in the real flood discharge space. As Table 5.1 shows, the expected discharge for the 100-year flood $(p = 0.01)$ is 4,310 cfs, the upper confidence limit is 6,176 cfs, and the lower limit is 3,008 cfs. The difference between the mean and the upper confidence limit is thus about 40 percent larger than the difference between the mean and the lower confidence limit. The confidence limits for graphical frequency analysis are computed using a method based on order statistics, as described in USACE (1997d). In this method, a given flood discharge estimate is considered a sample from a binomial distribution, whose parameters p and n are the nonexceedance probability of the flood and the equivalent record length of flood observations in the area, respectively. In this case, $n = 56$ years, since this is the record length of the Trevallian Way gage.

River Hydraulics

Water surface profiles for all events were determined using the HEC-2 river hydraulics program from the Corps's Hydrologic Engineering Center in Davis, California. Field-surveyed cross sections were obtained

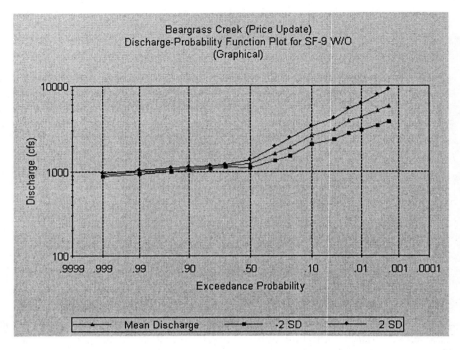

FIGURE 5.4 The flood–frequency curve and its uncertainty at damage reach SF-9 on the South Fork of Beargrass Creek.

at all bridges and at some stream sections near bridges. Maps with a scale of 1 inch = 100 feet with contour intervals of 2 feet were used to define cross sections elsewhere on the stream reaches and were used for measuring the distance between cross sections on the channel and in the left and right overbank areas. Manning's *n* values for roughness were based on field inspection, on reproduction of known high-water marks from the March 1964 flood on Beargrass Creek, and on reproduction of the rating curve of the USGS gage at Trevallian Way. Manning's equation relates the channel velocity to the channel's shape, slope, and roughness. Manning's *n* is a numerical value describing the channel roughness. Manning's *n* values in the concrete channel ranged from 0.015 at the channel invert to 0.027 near the top of the bank. In the natural channels, Manning's *n* values ranged from 0.035 to 0.050. In the overbank areas, these values ranged from 0.045 to 0.065. Where buildings blocked the flow, the cross sections were cut off at the effective

TABLE 5.1 Uncertainties in Estimated Discharge Values at Reach SF-9

Exceedance Probability	Mean Discharge (cfs)	Mean −2 Std Dev. (cfs)	Mean + 2 Std Dev. (cfs)	$\sigma_{\log Q}$
0.01	4,310	3,008	6,176	0.0781
0.1	2,620	2,051	3,346	0.0531
0.5	1,220	1,098	1,356	0.0229

flow limits. A total of 201 cross sections were used for the South Fork of Beargrass Creek, and 61 cross sections were used for Buechel Branch. The average distance between cross sections was 330 feet on the South Fork of Beargrass Creek and 245 feet on Buechel Branch. Cross sections are spaced more closely than this near bridges and more sparsely in reaches where the cross section is relatively constant.

Figure 5.5 shows the water surface profiles along Beargrass Creek for the eight flood frequencies considered, under existing conditions without any planned control measures. The horizontal axis of this graph is the distance in miles upstream from Beargrass Creek's outlet on the Ohio River. The vertical axis is the elevation of the water surface in feet above mean sea level. The bottom profile in this graph is the channel invert or channel bottom elevation. The top profile is for $p = 0.002$—the 500-year flood. This particular profile shows a sharp drop near the bottom end of the channel, caused by a bridge at that location that constricts the flow. The flat water surface elevation upstream of the bridge is a backwater effect produced by the inadequate capacity of the bridge opening to convey the flow that comes to it.

For each flood profile computed, the number of structures flooded and the degree to which they are flooded must be assessed. Figure 5.6 shows the locations of the first-floor elevations of structures affected by flooding on the South Fork of Beargrass Creek in relation to several flood water surface profiles under existing conditions. Damage reach SF-9 is located between river miles (RM) 9.960 and 10.363, near the point where there is a sharp drop in the channel bed and water surface elevation on Beargrass Creek. It can be seen that the density of development varies along the channel. Flood damage reduction measures are most effective when they are located close to damage reaches with significant numbers of structures, and they are least effective when they are distant from such reaches.

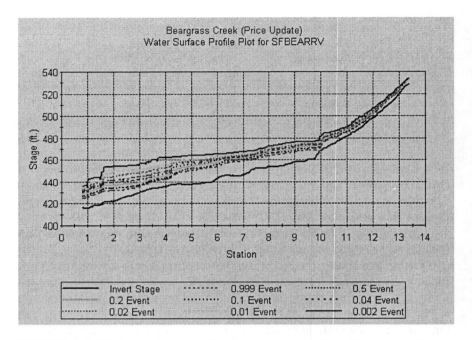

FIGURE 5.5 Water surface profiles for design floods in Beargrass Creek under existing conditions.

Each damage reach has an index location, which is an equivalent point at which all of the damages along the reach are assumed to occur. On reach SF-9, this index location is at river mile 10.124. To assess damages to structures within each reach, an equivalent elevation is found for each structure at the index location such that its depth of flooding at that location is the same as it would have been at the correct location on the flood profile, as shown in Figure 5.7.

The technique of assigning an elevation at the index location can be far more complex than Figure 5.7 implies, because allowance is made in the HEC-FDA program for the various flood profiles to be nonparallel and also to change in gradient upstream of the index location compared to downstream. In the Beargrass Creek study, a single flood profile for the $p = 0.01$ event was chosen, and all other profiles were assumed parallel to this one. One damage reach on Beargrass Creek was subdivided into three subreaches to make this assumption more nearly correct. A spatial distribution of buildings over the damage reach is thus converted

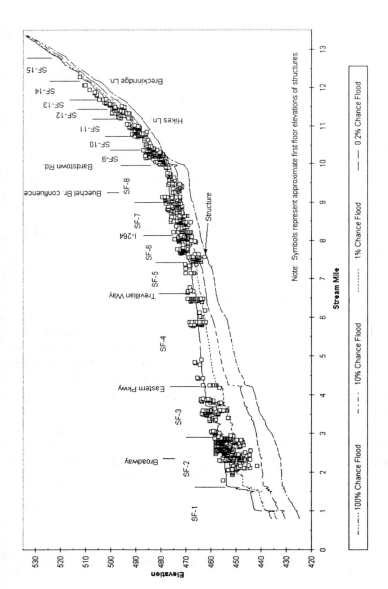

FIGURE 5.6 Locations of structures on floodwater surface profiles along the damage reaches of the South Fork of Beargrass Creek. SOURCE: USACE, 1997c.

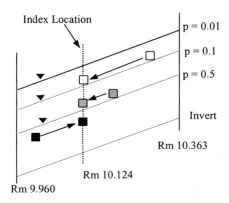

FIGURE 5.7 Assignment of structures to an index location.

into a probability distribution of buildings at the index location, where the uncertainty in flood stage is quantified.

Uncertainty in Flood Stage

The uncertainty in the water surface elevation was quantified by assuming that the standard deviation of the elevation at the index location for the 100-year discharge is 0.5 feet. The 100-year discharge at reach SF-9 is 4,310 cfs, which is the next to last set of points in Fugure 5.8. To the right of these points, between the 100-year and 500-year flood discharges, the uncertainties are assumed to be constant. For discharges lower than the 100-year return period, the uncertainties in stage height are reduced linearly in proportion to the depth of water in the channel. The various lines shown in Figure 5.8 are drawn as the expected water surface elevation ± 1 or 2 standard deviations determined in this manner.

Economic Analysis

The Corps's analysis of a flood damage reduction project's economic costs and benefits is guided by the *Principles and Guidelines* (Box 1.1 provides details on the *P&G*'s application to flood damage reduction

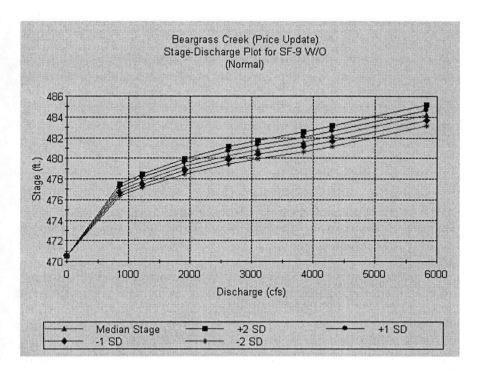

FIGURE 5.8 Uncertainty in the flood stage for existing conditions at reach SF-9 of the South Fork of Beargrass Creek.

studies). According to the *P&G*, the economic analysis of damages avoided to floodplain structures because of a flood damage reduction project is restricted to *existing* structures (i.e., federal policy does not allow damages avoided to prospective future structures to be counted as benefits). The *P&G* do, however, call for the benefits of increased net income generated by floodplain activities after a project has been constructed (so-called "intensification benefits") to be included in the economic analysis.

Economic analysis of flood damages considers various sorts of flood damage, principal among them being the damage to flooded structures. Information about the structures is quantified using a "structure inventory," an exhaustive tabulation of every building and other kind of structure subjected to flooding in the study region. A separate computer program called Structure Inventory for Damage Analysis (SID) was used to evaluate the number of structures flooded as a function of water surface elevation. Structures are divided into four categories: single-family residential, multifamily residential, commercial, and public. A structure is considered to be flooded if the computed flood elevation is above its first-floor elevation. The amount of damage D is a function of the depth of flooding h and the type of structure, and is expressed by a factor, $r(h)$, which is equal to a percentage of the value of the structure (V) and of its contents (C). This analysis can be expressed as

$$D = r_1(h)V + r_2(h)C . \tag{5.2}$$

For residential structures, these damage factors were quantified in 1995 by the Federal Emergency Management Agency (FEMA) using data from flood damage claims. For example, for a one-story house without a basement flooded to a depth of 3 feet, the FEMA estimate is that the damage factors are $r_1 = 27\%$ of the value of the structure and $r_2 = 35\%$ of the value of the contents. For the same house flooded to a depth of 6 feet, the corresponding damage factors are $r_1 = 40\%$ for the structure, and $r_2 = 45\%$ for the contents, respectively. The Marshall and Swift Residential Cost Handbook (Marshall and Swift, 1999) was used to estimate the value of single- and multi-family structures (it bears mentioning that the use of standard references such as the Marshall and Swift handbook may potentially represent another source of "knowledge uncertainty"). The values of their contents were assumed to be 40 percent to 44 percent of the value of the structure. For commercial and public buildings, the values of the structures and their contents were established through personal interviews by Corps personnel. About 85 percent of the structures subject to flood damage are residential buildings.

Types of flood damages beyond those to structures were also considered. For instance, there are several automobile sales lots in the floodplain, and prospective damages to cars parked there during a flood were estimated. Nonphysical damage costs include the costs of emergency services and traffic diversion during flooding. Damage to roads and utilities were also considered.

Uncertainty in Flood Damage

The economic analysis has three sources of uncertainty:

- the elevation of the first floor of the building,
- the degree of damage given the depth of flooding within the building, and
- the economic value of the structure and its contents.

For most structures in Beargrass Creek, the first-floor elevation was estimated from the ground elevation on maps with a scale of 1 inch = 100 feet and with contour intervals of 2 feet. For a sample of 195 structures (16% of the total number), the first-floor elevations were surveyed. It was found that the average difference between estimated and surveyed first-floor elevations of these structures was 0.62 feet.

Corps Engineering Manual (EM) 1110-2-1619 (USACE, 1996b) was used to estimate values for the uncertainties in economic analysis. A standard deviation of 0.2 feet was used to define the uncertainty in first-floor elevations. The uncertainty in the degree of damage given a depth of inundation was estimated by varying the percent damage factor described previously. For residential structures the value of the structure was assigned a standard deviation of 10 percent of the building value, and the ratio of the value of the contents to the structure was allowed to vary with a standard deviation of 20 percent to 25 percent.

For commercial property a separate damage estimate, based on interviews with the owners, was made for each significant property and was expressed as a triangular distribution with a minimum, expected, and maximum damage value for the property. Because every individual structure potentially affected by flooding is inventoried in the damage estimate data, the amount of work required to collect all these damage data was extensive.

The end result of these estimates at each damage reach and damage category is a damage–stage curve (such as Figure 5.9) that accumulates the damage to all multifamily structures in this damage reach for various water surface elevations at the index location, denoted by stage on the horizontal axis. This curve is prepared by first dividing the range of the stage (476–486 feet) into increments—increments of 0.5 feet in this case. For each structure, a cycle of 100 Monte Carlo simulations is carried out in which the first-floor elevation and the values of the structure and contents are randomly varied. From these simulations estimates are formed for each 0.5-foot stage height increment of what the expected damage and standard deviation of the damage to that structure would be if the flood stage were to rise to that elevation. For each stage increment, these means and standard deviations are accumulated over all structures in the reach to form the estimate of the mean and standard deviation of the reach damage (Figure 5.9).

A similar function is prepared for each of the damage categories. At any flood stage, the sum of the damages across all categories is the total flood damage for that reach.

Project Planning

The discussion of the Beargrass Creek study reviewed the technical means by which a particular flood damage reduction plan is evaluated. A plan consists of a set of flood damage reduction measures, such as detention ponds, levees or floodwalls, and channel modifications, implemented at particular locations on the creek. The base plan against which all others are considered is the "without plan," which means a plan that considers existing conditions in the floodplain and the development expected to occur even in the absence of a flood damage reduction plan. Such development must meet floodplain management policies and have structures elevated out of the 100-year floodplain. A base year of 1996 was chosen for the Beargrass Creek study.

In carrying out project planning, the spatial location of the principal damage reaches is important because flood damage reduction measures located just upstream of or within such reaches have greater economic impact than do flood damage reduction measures located in areas of low flood damage. Project planning also involves a great deal of interaction with local and state agencies, in this case principally the Jefferson County Metropolitan Sewer District.

The Beargrass Creek project planning team consisted primarily of three individuals in the Corps's Louisville district office: a project planner from the planning division, a hydraulic engineer from the hydrology and hydraulics design section, and an economic analyst from the economics branch. The HEC-FDA computer program with risk analysis was carried out by the economic analyst using flood–frequency curves and water surface profiles supplied by the hydrology and hydraulics section and using project alternatives defined by the project planner. The hydrology and hydraulics section was also responsible for the preliminary sizing of potential project structures being considered as plan components. The bulk of the work of implementing the risk analysis aspects of flood damage assessment thus fell within the domain of the Corps economic analyst.

The HEC-FDA program is applied during the feasibility phase of

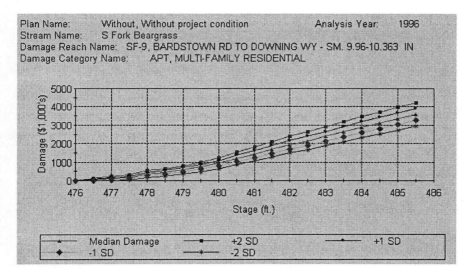

FIGURE 5.9 The damage–stage curve with uncertainty for multifamily residential property in Reach SF-9 of the South Fork of Beargrass Creek.

flood damage reduction planning. This had been preceded by a recon-naissance phase, a preliminary assessment of whether reasonable flood damage reduction planning can be done in the area. As explained in Chapter 2, the reconnaissance phase is fully funded by the federal gov-ernment, but the feasibility phase must have half the costs met by a local sponsor. Assuming the feasibility phase yields an acceptable plan and additional funds are authorized, the project proceeds to a detailed design and construction phase, which also requires local cost sharing. The Beargrass Creek project is now (as of May 2000) in the detailed design phase.

Evaluation of Project Alternatives

Expected annual flood damages in Beargrass Creek under existing conditions are estimated to be $3 million. Project benefits are calculated as the difference between this figure and the lower expected annual dam-ages that result with project components in place. Project costs are an-nualized values of construction costs discounted over a 50-year period using an interest rate of 7.625 percent. Project net benefits are the difference between project benefits and costs. For components to be included in the project, they must have positive net benefits.

The first step in evaluating project alternatives is to consider each component flood damage reduction measure by itself to see if it yields positive net benefits. A total of 22 components were examined individually, 11 on the South Fork of Beargrass Creek and 11 on Buechel Branch. All 11 of the South Fork components were economically justified on a stand-alone basis. Only 3 of the 11 components on Buechel Branch were justified individually: the other 8 components were thus deleted from further consideration.

The next step is to formulate the National Economic Development (NED) plan. In theory, this is supposed to proceed by selecting first the component with the largest net benefits, adding the component with the next largest net benefits, evaluating them together, and continuing to add more components until the combined set of components has the largest overall net benefits. It turned out that this idealized approach could not be used at the South Fork of Beargrass Creek because of economic and hydraulic interactions among the components. The study team commented: "Therefore, the formulation process was different and more complicated than originally anticipated. The study team could not follow the incremental analysis procedure to build up the NED plan because the process became a loop of H&H computer runs. Our component with the greatest net benefits is located near the midpoint of the stream; thus, each time we would add a component upstream it would affect all components downstream and vice versa. We could never truly optimize or identify the plan which produces the greatest net benefits" (USACE, 1997c, p. IV-62).

The problems were further complicated by the fact that there are three separate sections of the study region: the South Fork of Beargrass Creek and Buechel Branch upstream of their junction and the South Fork downstream of this junction (Figure 5.3). In the downstream region, flood damage reduction measures on the upper South Fork and Buechel Branch compete for project benefits by reducing flood damages. The result of these complications is that the plan was built up incrementally by separately considering the three sections of the region. First, the most upstream control structure in each section was selected, then structures downstream were added. At the end—when the components from the three sections had been aggregated into a single overall plan—it was determined whether the plan could be improved by omitting individual marginal components. The end result of this iterative process was a recommended plan with 10 components: 8 detention basins, 1 floodwall, and 1 channel improvement.

Each plan has to be evaluated using the Monte Carlo simulation process. The number of simulations varies by reach, with 10,000 required for Reach SF-9 and with a range of 10,000–100,000 required for the other reaches. On a 300 MHz Pentium computer, evaluation of a single plan takes about 25 minutes of computation time.

Risk of Flooding

The HEC-FDA program also produces a set of statistics that quantify the risk of being flooded in any reach for a given plan, as shown in Table 5.2. For reach SF-9, the target elevation is 477.2 feet, which is the elevation of the overbank area in this reach. The probability estimates shown are annual exceedance probability and conditional nonexceedance probability. The *annual* exceedance probability refers to the risk that flooding *will occur* considering all possible floods in any year. The *conditional* nonexceedance probability describes the likelihood that flooding *will not occur* during a flood of defined severity, such as the 100-year (1 percent chance) flood.

There is a subtle but important distinction between these two types of risk measures. The annual exceedance probability accumulates all the uncertainties into a single estimate both from the natural variability of the unknown severity of floods and from the knowledge uncertainty in estimating methods and computational parameters. The conditional nonexceedance probability estimate divides these two uncertainties, because it is conditional on the severity of the natural event and thus represents only the knowledge uncertainty component. In this sense, the conditional nonexceedance probability corresponds most closely to the traditional idea of adding 1 foot or 3 feet on the 100-year base flood elevation, while the annual exceedance probability corresponds more closely to the goal of ensuring that the chance of being flooded is less than a given value, such as 1 percent, considering all sources of uncertainty.

The "target stage annual exceedance probability" values in Table 5.2 are the median and the expected value or mean of the chance that flooding will occur in any given year for the various reaches. Thus, for reach SF-9, there is approximately a 36 percent chance that flooding will occur beyond the target stage in any given year, while in reach SF-14 upstream, that chance is only about 9 percent. The "long term risk" values in the

TABLE 5.2 Risk of Flooding in Damage Reaches Calculated Uncertainty for 1996 at Beargrass Creek

Damage Reach	Damage Reach Description	Target Stage (ft)	Target Stage Annual Exceedance Probability		Long-Term Risk (years)	
			Median	Expected	10	25
SF-8	Bashford Manor Lane to Bardstown Road	470.8	0.0920	0.0920	0.6182	0.9099
SF-9	Bardstown Road to Downing Way	477.2	0.3570	0.3640	0.9892	1.0000
SF-10	Downing Way to Goldsmith Lane	482.2	0.4240	0.4550	0.9977	1.0000
SF-11	Goldsmith Lane to Hikes Lane	487.7	0.0720	0.0780	0.5566	0.8691
SF-12	Hikes Lane to Klondike Lane	491.6	0.1220	0.1230	0.7310	0.9625
SF-13	Klondike Lane to Mid Dale Lane	495.1	0.1220	0.1200	0.7230	0.9596
SF-14	Mid Dale Lane to Breckin-ridge Lane	500.8	0.0990	0.0950	0.6237	0.9182
SF-15	Breckin-ridge Lane to Hunsinger	509.4	0.1030	0.1010	0.6561	0.9306

Note: Without project base year performance target criteria: event exceedance probability = 0.04 and residual damage = 5.00%.

Conditional Exceedance Probability by Events						
50	10%	4%	2%	1%	.4%	.2%
0.9919	0.5686	0.1939	0.0146	0.0027	0.0009	0.0005
1.0000	0.0027	0.0006	0.0000	0.0000	0.0000	0.0000
1.0000	0.0004	0.0000	0.0000	0.0000	0.0000	0.0000
0.9829	0.6489	0.3206	0.0903	0.0317	0.0082	0.0031
0.9986	0.3106	0.0718	0.0078	0.0011	0.0004	0.0001
0.9984	0.3127	0.0733	0.0093	0.0021	0.0004	0.0002
0.9933	0.5190	0.1978	0.0384	0.0100	0.0020	0.0008
0.9952	0.4919	0.1361	0.0294	0.0092	0.0029	0.0014

figure refer to the chance (R_n) that there will be flooding above the target stage at least once in n years, determined by the formula

$$R_n = 1 - (1-p_e)^n,\qquad\qquad\qquad (5.3)$$

where p_e is the expected annual exceedance probability. For example, for reach SF-9, where $p_e = 0.3640$, for $n = 10$ years, $R_{10} = 1 - (1 - 0.3640)^{10} = 0.9892$, as shown in Table 5.2.

The conditional nonexceedance probability values shown on the right-hand side of Table 5.2 are conditional risk values that correspond to the reliability that particular floods can be conveyed without causing damage in this reach. Thus, in reach SF-9, a 10 percent chance event (10-year flood) has about a 0.27 percent chance of being conveyed without exceeding the target stage, while for a 1 percent chance event (100-year flood), there is essentially no chance that it will pass without exceeding the target stage. By contrast, in Reach SF-14 at the upstream end of the study area, the conditional nonexceedance probability of the reach passing the 10-year flood is about 52 percent; that of the reach passing the 100-year flood is about 100 percent. As the flood severity increases, the chance of a reach being passed without flooding diminishes.

Effect on Project Economics of Including Risk and Uncertainty

The HEC-FDA program that includes risk and uncertainty factors in project analysis became available to the Beargrass creek project team late in the study period. Before then, the team used an earlier economic analysis program (Expected Annual Damage, or EAD) which computed expected annual damages without these uncertainties. O'Leary (1997) presented the data shown in Table 5.3 to compare the two approaches. It is evident that including risk and uncertainty increases the expected annual damage both with and without flood damage reduction plans. The net effect of their inclusion on the Beargrass Creek project is to increase the annual flood damage reduction benefits from $2.078 million to $2.314 million. The study team made a comparison between the components included in the National Economic Development plan in the two computer programs and found that there was no change. Hence, although the inclusion of risk and uncertainty increased project benefits, it did not result in changing the flood damage reduction components included in the National Economic Development plan.

O'Leary (1997) also presented statistics of the project benefits derived from the HEC-FDA program for the National Economic Development plan. The expected annual benefits of the National Economic Development plan—$2.314 million—are the same in Tables 5.3 and 5.4. The net benefits in the fourth column of Table 5.4 are found by subtracting the annual project costs from the expected annual benefits; the benefit-to-cost ratio is the ratio of the expected benefits to costs.

The 25th percentile, median (50th percentile), and 75th percentile of the expected annual benefits are also shown. The project net benefits are positive at all levels of assessment, and all benefit-to-cost ratios are greater than 1.00. It is interesting to see that the median expected annual benefits ($2.071 million) are nearly the same as the expected value of these benefits without considering uncertainty ($2.078 million). Moreover, the expected value ($2.314 million) is greater than the median, and the difference between the 75th percentile and the median is greater than the difference between the median and the 25th percentile. All these characteristics point to the fact that the distributions of flood damages and of expected annual benefits are positively skewed when uncertainties in project hydrology, hydraulics, and economics are considered. This is why the project benefits increase when these uncertainties are considered. The project benefits for the 25th percentile, 50th percentile, and 75th percentile in Table 5.4 should be read with caution because they are compiled for the project by adding together the corresponding values for all the damage reaches. The percentile value of a sum of random variables is not necessarily equal to the sum of the percentile values of each variable.

TABLE 5.3   Expected Annual Damages (EAD) With and Without Uncertainty in Damage Computations (millions of dollars per year)

Analysis Program	EAD Without Plan	EAD With NED Plan	Expected Annual Benefits
EAD Program (no uncertainty)	3.015	0.937	2.078
HEC-FDA (considering uncertainty)	3.998	1.684	2.314

SOURCE: O'Leary (1997).

TABLE 5.4   Statistics of project benefits under the NED plan using the HEC-FDA Program

Statistic	Expected Annual Benefits	Annual Project Costs	Net Benefits	Benefit to Cost Ratio
Expected Mean	2.314	0.810	1.504	2.86
25th Percentile	1.365	0.810	0.555	1.69
Median (50%)	2.071	0.810	1.261	2.56
75th Percentile	3.054	0.810	2.244	3.77

SOURCE: O'Leary (1997).

RED RIVER OF THE NORTH AT EAST GRAND FORKS, MINNESOTA, AND GRAND FORKS, NORTH DAKOTA

A devastating flood occurred at East Grand Forks, Minnesota, and Grand Forks, North Dakota, in April 1997. After the flood, flood damage reduction studies previously done for the two cities were combined into a joint study, and risk analysis was performed to evaluate the reliability of the proposed alternatives and to evaluate their economic impacts. A risk analysis study performed before the flood was presented in a paper at the Corps's 1997 Pacific Grove, California, workshop (Lesher and Foley, 1997). This paper and subsequent analysis (USACE, 1998a, b, c), as well as a visit to the Corps's St. Paul district office by a member of this committee, form the basis of this discussion of the East Grand Forks–Grand Forks study.

East Grand Forks, Minnesota, and Grand Forks, North Dakota, are located on opposite banks of the Red River of the North and are approximately 300 miles above the river's mouth at Lake Winnipeg, Manitoba, Canada (Figure 5.10). The East Grand Forks–Grand Forks metropolitan area has a population of approximately 60,000 and is located about 100 miles south of the U.S.–Canadian border. The total drainage area of the East Grand Forks–Grand Forks basin is 30,100 square miles. Included in this drainage area is the Red Lake River sub-basin that effectively drains about 3,700 square miles in Minnesota and joins the mainstream of the Red River at East Grand Forks. The study area of East Grand Forks–Grand Forks lies in the middle of the Red

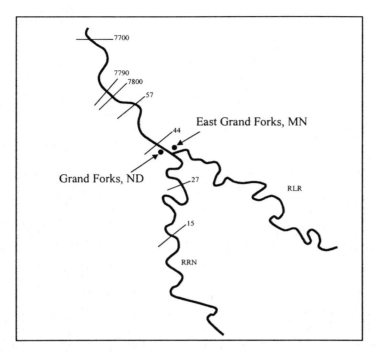

FIGURE 5.10 Schematic of the Red River of the North (RRN) and Red Lake River (RLR) at the East Grand Forks, Minnesota and Grand Forks, North Dakota study area. Numbers indicate USGS stream gages.

River Valley. The valley is exceptionally flat with a gradient that slopes 3–10 feet per mile toward the river with the north–south axis having a gradient of about three-quarters of a foot per mile. The valley extends approximately 23 miles west and 35 miles east of East Grand Forks–Grand Forks and is a former glacial lake bed.

Both cities have a long history of significant flooding from the Red River of the North and the Red Lake River. The most damaging flood of record occurred in April 1997 (see Table 5.5), when the temporary levee systems and flood-fighting efforts of both communities could not hold back the floodwaters of the Red River. The resulting damages were disastrous and affected both cities dramatically. Total damages to existing structures and contents during the 1997 flood were estimated to exceed $800 million. An additional $240 million was spent for emergency-related costs.

TABLE 5.5 Maximum Recorded Instantaneous Peak Flows;
Red River of the North at Grant Forks, North Dakota

Order	Date	Discharge (cfs)
1	April 18, 1997	136,900
2	April 10, 1897	85,000
3	April 26, 1979	82,000
4	April 18, 1882	75,000
5	April 21, 1996	58,400
6	April 4, 1966	55,000
7	April 11, 1978	54,200
8	May 12, 1950	54,000
9	April 16, 1969	53,500
10	April 24, 1893	53,300

SOURCE: USACE (1998a).

Risk Analysis

A risk analysis for the proposed flood damage reduction project for the Red River of the North at East Grand Forks, Minnesota, and Grand Forks, North Dakota, used a Latin Hypercube analysis to sample interactions among uncertain relationships associated with flood discharge and elevation estimation. Latin Hypercube is a stratified sampling technique used in simulation modeling. Stratified sampling techniques, as opposed to Monte Carlo-type techniques, tend to force convergence of a sampled distribution in fewer samples. Because the Hydrologic Engineering Center Flood Damage Analysis program (HEC-FDA) was new at the time, and in the interest of saving time, the analysis was performed using a spreadsheet template. The flood damage reduction alternatives analyzed included levees of various heights and a diversion channel in conjunction with levees. The project reliability option in the HEC risk spreadsheet was used to determine the reliability of the alternative levee heights and of the diversion channel in conjunction with levees. The following sections discuss the sensitivity in quantifying the uncertainties and the representation of risk for the alternatives.

Discharge–Frequency Relationships

The log-Pearson Type III distribution, recommended in the Water Resource Council's *Bulletin 17B* (IACWD, 1981) and incorporated within the Corps's HEC Flood Frequency Analysis (HEC-FFA) computer program, was used for frequency analysis of maximum annual streamflows, and the noncentral t distribution was used for the development of confidence limits. Discharge–frequency relationships were needed for both the levees and the diversion channel in combination with levees. An analysis (coincidental frequency) was performed to develop the discharge–frequency curves for the Red River of the North downstream and upstream of the Red Lake River for the levees only condition. A graphical method was used to develop the discharge–frequency curves for the diversion channel in combination with levees. Details of these procedures can be found in a Corps instruction manual from the St. Paul district (USACE, 1998a). A brief discussion of these procedures is provided below.

The Grand Forks USGS stream gage (XS 44) is currently located 0.4 miles downstream from the Red Lake River in Grand Forks, North Dakota (Figure 5.10). The discharge–frequency curve for this station along with the 95 percent and 5 percent confidence limits (90% confidence band) are plotted in Figure 5.11. An illustration of the noncentral t probability density function for the 1 percent event is also shown in that figure. Selected quantities of that discharge–frequency relationship are shown in column 2 of Table 5.6. The coincidental discharge–frequency relationship for the Red River just upstream of the mouth of the Red Lake River (column 3 of Table 5.6) was computed with the HEC-FFA computer program. The basic flow values were obtained by routing the 96 years of available data on Red Lake River flows from Crookston (55 miles upstream of the mouth) downstream to Grand Forks. The resulting flows were subtracted from the Red River at Grand Forks flows to obtain coincident discharges on the Red River upstream of Red Lake River. The two-station comparison method of *Bulletin 17B* was used to adjust the logarithmic mean and standard deviation of this short record (96 years) based on regression analysis with the long-term record at the Grand Forks station (172 years). Correlation of coincident flows for the short record with concurrent peak flows for the long record produced a correlation coefficient of 0.975.

Adjustment of the statistics yielded an equivalent record length of 165 years. The adopted coincidental discharge–frequency curve for the Red River upstream of the Red Lake River is shown in column 3 of Table 5.6 for selected annual exceedance probabilities. The coincidental discharge–frequency curve for the Red Lake River at the mouth was determined by computing the difference in Red River flows both upstream

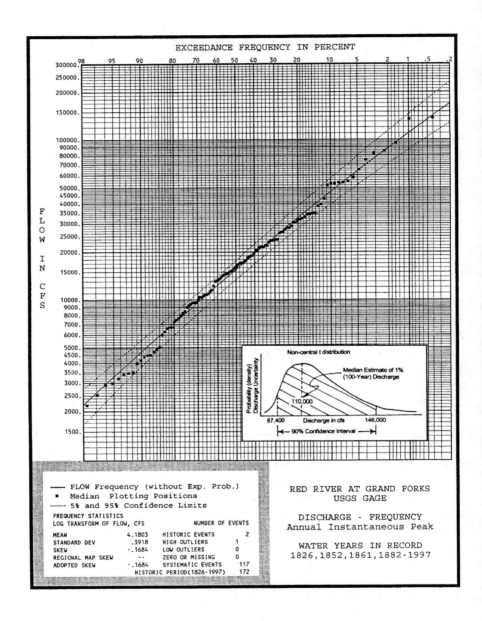

FIGURE 5.11   Flood (discharge) frequency curve for the Red River at Grand Forks.

TABLE 5.6   Instantaneous Annual Peak Discharges (cfs) and their Annual Exceedance Probabilities (%) — Existing Conditions

	Instantaneous Annual Peak Discharges (cfs)		
Annual Exceedance Probability in Percent	Red River of the North Below Red Lake River	Red River of the North Above Red Lake River	Red Lake River at the Mouth [based on difference]
0.2	169,000	128,000	41,000
0.47	136,900	102,000	34,900
0.5	134,000	100,000	34,000
1.0	110,000	81,700	28,300
2.0	89,000	64,900	24,100
5.0	63,900	45,500	18,400
10.0	47,300	32,900	14,400
20.0	32,600	21,900	10,700
50.0	15,500	9,590	5,910
80.0	7,150	3,970	3,180
90.0	4,700	2,450	2,250
95.0	3,290	1,620	1,670
99.0	1,660	726	934

and downstream of Red Lake River (see column 4 in Table 5.6). Statistics for the adopted relationship were approximated by synthetic methods presented in *Bulletin 17B* (for more details, see USACE (1998a)).

The Plan Comparison Letter Report developed in February 1998 for flood damage reduction studies for East Grand Forks, Minnesota, and Grand Forks, North Dakota, evaluated an alternative flood damage reduction plan that included a split-flow diversion channel along with permanent levees. The discharge–frequency relationships for the modified conditions, shown in Table 5.7, were developed as follows. The modified-condition discharge–frequency curve for the Red River upstream of Red Lake River was graphically developed based upon the operation of the diversion channel inlet. Red River flows are not diverted until floods start to exceed those having return periods of 5 years (20% annual exceedance probability). The channel is designed to continue to divert Red River flows at a rate that allows the design flood (0.47%) discharge of 102,000 cfs (upstream of the diversion) to be split such that 50,500 cfs is diverted and 51,500 cfs is passed through the cities. This operation is reflected in the modified discharge–frequency relationship shown in Table 5.7 for the Red River upstream of Red Lake River (columns 2 and

TABLE 5.7   Instantaneous Annual Peak Discharges (cfs) and their Annual Exceedance Probabilities (%)—Condition with Diversion Channels

Annual Exceedence Probability in Percent	Instantaneous Annual Peak Discharges (cfs) Red River of the North Above Red Lake River		Red Lake River at the Mouth	Red River of the North Below Red Lake River
	Above Diversion	Below Diversion		
0.2	128,000	55,000	41,000	96,000
0.47	102,000	51,500	34,900	86,400
0.5	100,000	51,000	34,000	85,000
1.0	81,700	47,500	28,300	75,800
2.0	64,900	43,000	24,100	67,100
5.0	45,500	36,500	18,400	54,900
10.0	32,900	30,000	14,400	44,400
20.0	21,900	21,900	10,700	32,600
50.0	9,590	9,590	5,910	15,500

3).Synthetic statistics (mean, standard deviation, and skewness) in accordance with methodology presented in *Bulletin 17B* were computed for the discharge-frequency relationships of the below-diversion flows.

The modified-condition discharge–frequency curve for the Red River downstream of Red Lake River was graphically computed based upon the operation of the diversion channel. The modified-condition Red River discharges upstream of Red River were added to the coincident flows on Red Lake River (column 4). The resulting discharges were plotted for graphical development of the modified-condition discharge–frequency relationship for the Red River downstream of Red Lake River and are summarized in Table 5.7 (column 5). Synthetic statistics for this discharge–frequency relationship were computed for use in the risk analysis.

Elevation–Discharge Relationships

The water surface elevations computed using the HEC-2 computer program are shown in Table 5.8 for three cross sections (7790, 7800, and 7922) corresponding to the previous USGS gage locations and for cross section 44, which corresponds to the current USGS gage location (see Figure 5.10 for the cross section locations). These computed water surface elevations (CWSE) were based on the expected discharge quantities from the coincidental frequency analysis performed in June 1994 for the Grand Forks Feasibility Study. These data were used to transfer observed elevations from previous USGS gage sites to the current site (cross section 44) at river mile 297.65, and they were used in determining the elevation–discharge uncertainty. The water surface profile analysis was performed using cross-sectional data obtained from field surveys. Data were also obtained from field surveys and from USGS topographic maps. The HEC-2 model was calibrated to the USGS stream gage data and to high-water marks for the 1969, 1975, 1978, 1979 and 1989 flood events throughout the study area. Note that these water surface elevations assume the existing East Grand Forks and Grand Forks emergency levees are effective. The levees were assumed effective because through extraordinary efforts, they have generally been effective for past floods with the exception of the 1997 flood.

Ratings at stream gage locations provide an opportunity to directly analyze elevation–discharge uncertainty. The measured data are used to derive the "best fit" elevation-discharge rating at the stream gage location, which generally represents the most reliable information available. In this study, the adopted rating curve for computing elevation uncertainty is based on the computed water surface elevations from the calibrated HEC-2 model shown in Table 5.8.

This adopted rating curve for cross section 44 at the current USGS gage is shown in Figure 5.12. Measurements at the gage location were used directly to assess the uncertainty of the elevation–discharge relationship. The normal distribution was used to describe the distribution of error from the "best-fit" elevation–discharge rating curve. The observed gage data (for the four cross sections presented in Table 5.8) were transferred to the current gage site at river mile 297.65 based on the gage location adjustments presented in Table 5.9, which were computed from the water surface elevations in Table 5.8. These adjustments were plotted against the corresponding discharge below the Red Lake River, and curves were developed to obtain adjustments for other discharges.

The deviations of the observed elevations from the fitted curve were used to estimate the uncertainty of the elevation–discharge rating curve shown in Figure 5.11. The deviations reflect the uncertainty in data values as a result of changes in flow regime, bed form, roughness/resistance to flow, and other factors inherent to flow in natural streams. Errors also

TABLE 5.8 Computed Water Surface Elevations of the Red River of the North at Grand Forks, North Dakota (units in feet above sea level)

Cross Section Number	River Mile	Minimum Channel Bottom in Feet	Floods					
			20% (5-year)	10% (10-year)	4% (25-year)	2% (50-year)	1% (100-year)	0.2% (500-year)
7790	295.70	773.15	817.20	821.70	825.00	827.30	829.60	834.80
7800	296.00	774.2	817.39	821.87	825.19	827.52	829.83	835.01
7922	297.55	774.60	818.26	822.74	826.27	828.83	831.58	837.25
44[a]	297.65	772.40	818.39	822.91	826.67	829.18	831.84	837.59

[a] Current Location of USGS gage.

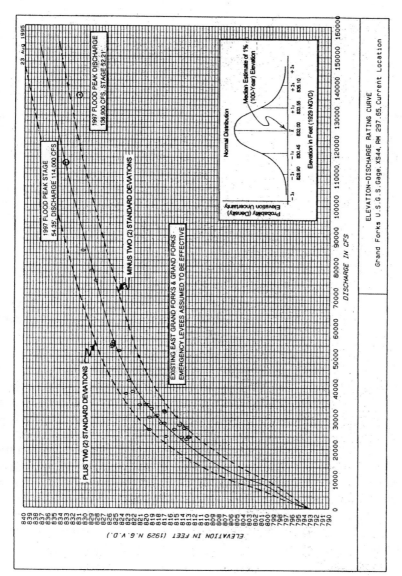

FIGURE 5.12 Rating curve (water elevation vs. discharge) for the Red River at Grand Forks.

TABLE 5.9 Adjustments Used in Transferring Observed Elevations from Previous USGS Gage Sites to Current Gage Site at RM 297.65 (XS 44)

	Expected Probability Discharge (cfs)		Adjustment Factor (cfs)		
Probability	Below Red Lake River	Above Red Lake River	XS 7790, RM 295.70	XS 7800, RM 296.00	XS 7922, RM 297.55
38%	20,000	12,500	1.28	1.06	0.12
27%	25,000	16,100	1.23	1.02	0.12
20%	30,600	20,300	1.19	1.00	0.13
10%	43,900	30,300	1.21	1.04	0.17
4%	63,500	45,800	1.67	1.48	0.40
2%	81,500	58,800	1.88	1.66	0.35
1%	101,000	73,500	2.24	2.02	0.26

result from field measurements or malfunctioning equipment. A minimum of 8–10 measurements is normally required for meaningful results. The measure used to define the elevation–discharge relationship uncertainty is the standard deviation:

$$\sigma = \sqrt{\frac{\Sigma(X - M)^2}{N - 1}}, \qquad (5.4)$$

Where X = observed elevation adjusted to current gage location (if 5.12 necessary), M = computed elevation from adopted rating curve, and N = number of measured discharge values (events).

The elevation uncertainty was computed for two different discharge ranges for this analysis. Based on the observed elevations plotted on the adopted rating curve, it appeared that there was greater uncertainty for discharges less than about 10% of annual exceedance probability event due to ice effects on flow. Therefore, the standard deviation was computed for discharges greater than between 22,000 cfs, which corresponds approximately to the zero damage elevation based on the adopted rating curve, and 44,000 cfs, which is slightly greater than the 10 percent annual exceedance probability. The standard deviation was also computed for discharges greater than 50,000 cfs. During the period of record, there were 25 events with a discharge between 22,000 and 44,000 cfs and 10 events with a discharge greater than 50,000 cfs. The standard deviation was 1.66 feet for discharges between 22,000 and 44,000 cfs and was 1.55

feet for discharges greater than 50,000 cfs. In the risk and uncertainty simulations, the standard deviation was linearly interpolated between 1.66 and 1.55 feet for discharges between 44,000 and 50,000 cfs. (See USACE (1998b) for more details.)

In an earlier risk analysis that was performed for the Grand Forks Feasibility Study, a much lower standard deviation of 0.50 feet was used for discharges greater than 50,000 cfs. However, adding the 1997 flood to the analysis resulted in a standard deviation of 1.55 feet, which is similar to that computed for discharges less than 44,000 cfs. It should be noted that the discharge and elevation used in this analysis for the 1997 flood was the peak discharge of 136,900 cfs occurring on April 18, 1997 (see Table 5.4), and an elevation of 831.21 feet (Stage 52.21). The peak elevation of 833.35 feet (Stage 54.35) occurred on April 22, 1997 at a discharge of 114,000 cfs. The elevation of 831.21 feet was almost 5 feet below the rating curve at a discharge of 136,900 cfs; however, the peak elevation of 833.35 feet at a discharge of 114,000 cfs was essentially on the adopted rating curve. Both of these points are plotted on the rating curve in Figure 5.12. Lines representing ± 2 standard deviations for the normal distribution, which encompasses approximately 95 percent of all possible outcomes, are also shown on the rating curve. An illustration of the normal distribution at the 1 percent (100-year) event for the project levee condition is also shown in Figure 5.12.

Risk and Uncertainty Analysis Results

Four index locations were selected to evaluate project performance and project sizing. These locations are cross sections 57, 44 (current USGS gage), 27, and 15 (Figure 5.10). The four locations were selected based on economic requirements for project sizing (see USACE, 1998c). The elevation–discharge rating curves (based on HEC-2 analysis) for existing and project conditions at these locations can be found in the USACE (1998b). Each of these rating curves shows three conditions, where applicable: (1) existing conditions, (2) removal of the pedestrian bridge at cross sections 7920-7922 and with project levees ("levee only"); and (3) with removal of the pedestrian bridge, with project levees, and with the diversion channel ("diversion channel"). Existing conditions means that the existing emergency levees are assumed to be effective up to and including the 5 percent (20-year) event and are ineffective for larger floods. The 5 percent (20-year) event was selected based on comparison of water surface profiles with effective and probable failure point (PFP) levee elevations provided by the Geotechnical Design Section analysis (see USACE, 1998b, paragraph A.2.11 and Appendix B of this report). The pedestrian bridge was removed based on input from the cities of East Grand Forks and Grand Forks. The rating curves for the diversion channel alternative were based on limited information. The Red River to the North would start to divert into the diversion channel at the 20 percent (5-year) flood; therefore, up to this point the rating curve for existing conditions with levees was used.

An additional location was also selected to evaluate the performance of the levee only and diversion channel with 1 percent (100-year) levee alternatives. This location is at cross section 7700 at the downstream end of the project levees (see Figure 5.10). Cross section 7700 was selected based on hydraulic analysis as the least critical location—the location where the levees in combination with the diversion channel would first overtop from downstream backwater (see USACE, 1998b).

Project Reliability

The project reliability results are summarized in Tables 5.10 through 5.12. Table 5.9 contains the results for the levees-only alternatives. Table 5.11 contains the results for the diversion channel in combination with 1 percent (100-year) levees. Note that in Table 5.10, three different alternative top-of-levee heights are evaluated, whereas in Table 5.11, it is always the same alternative—diversion channel with 1 percent levees— but for the three different events. The top-of-levee elevations were computed based on a water surface elevation profile to ensure initial overtopping would occur at the least-critical location (here, cross section 7700). The downstream top-of-levee elevations were selected with the intent of having 90 percent probability of containing the specified flood and were based on previous risk analysis for the Grand Forks Feasibility Study preliminarily updated to include the 1997 flood. The 2 percent (50-year), 1 percent (100-year), and 0.47 percent (210-year/1997 flood) top-of-levee profiles are 3.2, 3.4, and 2.7 feet above their respective water surface profiles at the downstream end (Table 5.10).

As seen in Table 5.10, the intent of having 90 percent probability of containing the specified flood is generally realized. The 2 percent levees have a 92 percent probability of containing the 2 percent flood. The 1 percent levees have a 90 percent probability of containing the 1 percent

TABLE 5.10 Reliability at Top of Levee for Three Top-of-Levee Heights

Gage Location	2 % (50-year) Levee[a]		1% (100-year) Levee[b]		0.47% (210-year) Levee[c]	
	Top of Levee (ft.)	Reliability (%)	Top of Levee (ft.)	Reliability (%)	Top of Levee (ft.)	Reliability (%)
XS 7700[d]	830.2	92.5	832.7	90.7	834.8	87.7
XS 57	832.0	92.0	834.2	90.5	836.2	86.4
XS 44	833.2	93.2	835.6	91.3	837.5	86.3
XS 27	834.3	92.1	836.9	89.5	839.0	86.5
XS 15	835.2	92.7	837.7	90.0	839.7	85.5

[a]Top of levee for the 2% levee is computed water surface elevation plus 3.2 feet.
[b]Top of levee for the 1% levee is computed water surface elevation plus 3.4 feet.
[c]Top of levee for 0.47% levee is computed water surface elevation plus 2.7 feet.
[d]Downstream end of project.

TABLE 5.11 Project Reliability at Top of Levee for Diversion Channel with 1 Percent (100-Year) Levees for Three Different Events

Gage Location	Top of Levee (ft)	Reliability 2% (50-year) Event	1% (100-year) Event	0.47% (210-year) Event
XS 7700[a]	832.7[b]	99.9	99.6	98.9
XS 57	834.2	100.0	99.6	99.2
XS 44	835.6	99.9	99.6	99.4
XS 27	836.9	99.6	99.5	99.1
XS 15	837.7	99.7	99.6	99.2

[a]Downstream end of project.
[b]Top of levee is computed water surface elevation plus 3.4 feet.

flood. The 0.47 percent levees have an 87 percent probability of containing the 0.47 percent flood.

Reliability results for the diversion channel with 1 percent levees are summarized in Table 5.11. Note again that the levees constructed in combination with the diversion are the same as for the 1 percent flood without the diversion channel and are the same for all three floods analyzed. As seen in the table, there is a 99 percent or greater probability of containing the flood for all three floods considered when the project includes the diversion channel.

As previously noted, the most critical location for project performance is at cross section 7700 at the downstream end of the project. Table

TABLE 5.12 Conditional Exceedance Probability of Alternative for Various Events (based on analysis at downstream end of project—XS 7700)

Event	Alternative 2% (50-year) Levees	1% (100-year) Levees	0.47% (210-year) Levees	Diversion with 1% (100-year) Levees
4 % (25-year)	99.5	100.0	100.0	100.0
2% (50-year)	92.5	99.1	99.7	99.9
1% (100-year)	64.3	90.7	98.3	99.6
0.52% (192-year)	29.5	65.6	89.8	—[a]
0.5% (200-year)	28.2	64.4	88.7	—[a]
0.47% (210-year)	25.3	61.9	87.7	98.9
0.2% (500-year)	4.4	21.5	48.0	>95
0.1% (1,000-year)	0.7	6.0	20.7	—[b]

[a]Event not analyzed.
[b]Event not analyzed because (1) the discharge–frequency curve based on the approximate statistics starts to diverge from the graphical curve for extreme events and (2) there was limited information to develop the RRN rating curves for the diversion alternative.

5.12 summarizes the results for all the alternatives considered and for numerous floods. The probability of the diversion channel in combination with 1 percent levees for the 0.2 percent event is listed in the table as greater than 95%. A more specific reliability was not cited for the 0.2 percent event for two reasons: (1) the discharge–frequency curve based on the approximate statistics starts to diverge from the graphical curve for extreme events and, (2) there was limited information available to develop the Red River to the North rating curves for the diversion alternative. These reasons are also why more extreme events were not analyzed.

Table 5.13 presents the simulated conditional exceedance probabilities from the economic project sizing analysis. The without-project condition is also included in this table for comparison purposes. The without-project condition is based on a zero damage elevation of 824.5 feet, assumes credit is given to the existing levees, and assumes all properties that were substantially damaged (50% or more damage) in the 1997 flood have been removed.

Based on the above analysis of alternative plans and further economic and environmental considerations, the recommended National

TABLE 5.13 Residual Risk Comparison

Alternative	Annual Performance (Expected Annual Probability of Design Being Exceeded)
Without Project	0.0918
2% (50-Year) Levees	0.0086
1% (100-Year) Levees	0.0036
0.47% (210-Year) Levees	0.0010
Diversion with 1% (100-Year) Levees	0.0002

Economic Development (NED) plan consists of a permanent levee and floodwall system designed to reliably contain the 210-year flood event. This equates to an 87.7 percent reliability of containing the 210-year flood event (Table 5.12) and would reliably protect against a flood of the magnitude of the 1997 flood.

The recommended plan would remove protected areas from the regulatory floodplain, increase recreational opportunities, and enhance the biological diversity in the open space created. The recommended plan anticipates the need to acquire over 250 single-family residential structures, 95 apartment or condominium units, and 16 businesses along the current levee/floodwall alignment.

The total cost of the recommended multipurpose project is $350 million including recreation features and cultural resources mitigation costs. The federal share of the project would be $176 million and the nonfederal share would be $174 million. The benefit-to-cost ratio has been calculated as 1.07 for the basic flood reduction features of the project and as 1.90 for the separable recreation features (USACE, 1998b). The recommended project has an overall benefit-to-cost ratio of 1.10.

The cities of East Grand Forks, Minnesota, and Grand Forks, North Dakota, will serve as the project's nonfederal sponsors. Through legislation, the State of Minnesota has committed to provide financial support in the form of bonds and returned sales taxes to the city of East Grand Forks. In verbal and written comments from its governor, the State of North Dakota has committed to provide financial assistance to the city of Grand Forks.

6

Evaluation and Proposed Improvements

Explicit recognition of modeling uncertainty in the Corps of Engineers's risk analysis methods should result in a better understanding of the accuracy of both flood risk and flood damage reduction estimates. Early applications of the risk analysis methods, however, illustrate that inadequacies must be overcome before the methods can be expected to yield consistent, defensible results.

This chapter recommends improvements the Corps should make in its risk analysis methods for flood damage reduction studies to overcome the shortcomings identified in previous chapters. This chapter addresses three broad issues: (1) the completeness of the set of uncertainties included in the analysis, (2) the importance of recognizing and effectively dealing with differences between natural variability and knowledge uncertainty, and (3) the treatment of interrelationships among the uncertainties.

CONCERNS WITH THE RISK ANALYSIS METHODS

The three key questions to be asked of risk analysis methods are:

- Is the set of uncertainties included in the analysis complete?
- Are uncertainties of different types treated appropriately?
- Are the interrelationships among uncertainties correctly quantified?

In addressing these questions, the current methods are only partially satisfactory. There are important ways in which the methods should be improved.

To some extent, improvements needed in the Corps's risk analysis methods arise from a lack of clearly articulated goals of the risk analysis. As noted in a 1995 National Research Council report (NRC, 1995, pp. 120–21, 136–43), "A framework is needed to understand the structure of risk and uncertainty analysis efforts for flood protection project evaluation, and to understand the relative roles of the natural variability of flood volumes, reservoir operations, hydraulic system performance, stage–discharge errors, and uncertainty in hydrologic, hydraulic, and economic parameters."

This recommendation remains relevant. The current framework does not span the full range of uncertainties important to flood risk and flood damage reduction, it does not clearly differentiate between natural variability and knowledge uncertainty or the relative roles of these uncertainties, and it does not recognize the importance of spatial structure and correlation among uncertainties. A clear a priori articulation of the goals of the risk analysis approach would help illuminate the needed structure of the analysis and help identify conceptual gaps.

ENGINEERING PERFORMANCE

The Corps's risk analysis methods in its flood damage reduction studies use probability concepts in several ways. These might be classified using three levels, as shown in Table 6.1. At the uppermost level, natural variability arising from inter-annual variations in flood severity is quantified by the flood–frequency curve and is indexed by a flood's annual probability of exceedance. At the middle level, project performance measures are defined using probabilities for engineering performance and dollars for economic performance. These measures incorporate both natural variability and knowledge uncertainty. At the lowest level, uncertainty in performance measures is specified. This is knowledge uncertainty in the output measures resulting from knowledge uncertainty in the input information. The input information pertains to hydrologic, hydraulic, geotechnical, and economic factors.

Knowledge Uncertainty

For economic performance, knowledge uncertainty is measured by percentile values of the expected annual damage and expected annual

TABLE 6.1 Levels of Inclusion of Probability in Risk Analysis

	Engineering Performance	Economic Performance
Natural Variability	Range of flood frequency	Range of flood frequency
Performance Measure	Annual exceedance probability, Conditional nonexceedance probability	Expected annual damage, Expected annual benefit
Uncertainty of the Performance Measure	No specification	Percentile values of annual damage and benefit
		Variance of expected annual damage, and of expected annual benefit

benefits of the project plan. A key omission in the Corps's risk analysis procedure is that there is no comparable quantification of knowledge uncertainty in the engineering performance. Variations in flood–frequency and stage–discharge relationships are defined based on uncertain knowledge of the analysis parameters, but there is no comparable specification of the resulting variation in the risk measures defining engineering performance.

In the previous chapter, Table 5.2 shows the results of a risk analysis calculation made using the Hydrologic Engineering Center Flood Damage Assessment (HEC-FDA). The table lists 11 probability measures of engineering performance, but each is a variation on the basic themes of annual exceedance probability or conditional nonexceedance probability. It is axiomatic that every measure of uncertainty is itself uncertain. Knowledge uncertainty in the economic performance measures is described by the percent variation in estimated values. There is presently no comparable presentation of knowledge uncertainty in the engineering performance measures reported by the analysis (with the exception that both the mean and the median of the annual exceedance probability are reported).

This omission had an important impact on the use of risk analysis in the procedures for levee certification. In 1993 the Federal Emergency Management Agency (FEMA) and the Corps agreed that a levee could be certified as providing adequate flood protection in the event the annual exceedance probability was less than 1 percent. In effect, they chose the mean value of this measure, ignoring uncertainties in the annual exceedance probability. After several years of this practice, some levees certified by this procedure provided inadequate levels of flood protection when compared to the previous criterion used (100-year flood plus 3 feet of freeboard), and the Corps and FEMA abandoned the 1 percent mean annual exceedance probability criterion for a more complicated scheme (described in Chapter 7). This policy change might not have been necessary had it been recognized from the beginning that the annual exceedance probability contains knowledge uncertainty, which can be quantified using percentile values like those used for the economic performance measures.

The Monte Carlo simulation upon which the risk analysis is based generates a set of N values of the annual exceedance probability of the target stage for each reach. Such a distribution computed for an arbitrary location in the Beargrass Creek study is shown in Figure 6.1. The vertical axis is the annual exceedance probability of the target stage at a damage reach. The horizontal axis is its cumulative probability distribution, which is found using the Monte Carlo simulation. The median value of the annual exceedance probability in this damage reach is 0.05, and the range is from 0.005 to 0.157, which measures the extent of the knowledge uncertainty of this statistic. The distribution is slightly positively skewed and the mean annual exceedance probability is 0.055. This means that in any year, the annual expected exceedance probability of flooding exceeding the target stage is about 5.5 percent, so the return period of flooding in this reach is approximately 18 years (1/0.055). This is the "chance of getting wet" for somebody living next to the creek in this reach.

Suppose one wished to define a conservative criterion limiting the "chance of getting wet." One such criterion could be the one-sided 90 percent confidence level value of the annual exceedance probability, shown in Figure 6.1 to be 0.113. This means that a person living next to the creek can be 90 percent sure that the chance of flooding in any given year is less than 11.3 percent. Equivalently, this citizen could be 90 percent sure that flooding would occur not more frequently than an average of once every 8.8 years (1/0.113). Thus, 8.8 years would be the level of

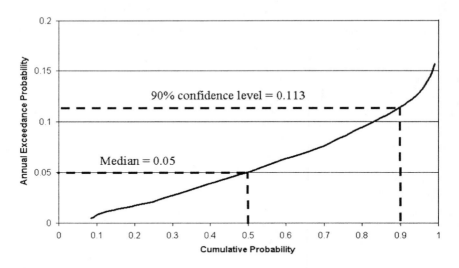

FIGURE 6.1. The distribution of the annual exceedance probability for a location in the Beargrass Creek study.

protection afforded to this citizen if the level of protection was defined to be the annual exceedance probability achieved with 90 percent confidence.

Engineering performance is currently measured by many different criteria (annual exceedance probability, both expected and median; estimated long-term risk for 10, 25, and 50 years; and conditional nonexceedance probability for $p = 0.1, 0.04, 0.02, 0.01, 0.004,$ and 0.002). The result is a range of risk measures in which the user can easily get lost. The concept of conditional nonexceedance probability is particularly confusing to the public, as pointed out by an earlier NRC committee (NRC, 1995, p. 162). By measuring engineering performance in a manner directly conformal to the way economic performance is measured, the effects of a project on both types of performance can be more readily stated and compared.

The committee recommends that the Corps standardize the annual exceedance probability as its principal engineering performance measure for decision making in flood damage reduction studies. The range of variation of this measure resulting from knowledge uncertainty should be specified by a table of percentile values in the same way as is currently done for the economic performance measures.

HYDROLOGIC ANALYSIS

Analysis of hydrologic and hydraulic uncertainties in the risk analysis method proceeds according to the interlinking relationships discussed in Chapter 3 and illustrated in Figure 3.1. This method is based on the Corps's traditional deterministic procedures and is validated by a long history of use. However, a number of specific improvements are needed, most of them reflecting inadequacies in the treatment of natural variability vs. knowledge uncertainty.

Parameter Uncertainty for the LP3 Distribution

The current algorithm described in Appendix F of the HEC-FDA manual for generating parameter uncertainty for the log-Pearson Type III (LP3) distribution first generates an uncertain mean and variance as if the logarithms of the floods are normally distributed. This can be done using the known sampling distribution for these parameters with normal samples. However, the algorithm then applies a transformation to generate the corresponding mean and standard deviation for an LP3 distribution, because the exact sampling distribution of the mean and variance of the LP3 distribution is not available. However, the approximation that is used, based upon an expected probability adjustment in *Bulletin 17B* (IACWD, 1981), has no theoretical justification of which the committee is aware. The sampling scheme implemented in HEC-FDA is intended to provide consistent estimates between traditional estimates of the expected probability frequency curve historically used by the Corps, and that derived from the Monte Carlo procedure. This allows consistent comparisons between risk analysis studies and the historical approach.

Additional uncertainty enters the calculation of flood damage due to knowledge uncertainty regarding the choice of model to use for the statistical distribution of floods. One way the Corps might attempt to address model uncertainty is by assessing whether the LP3 distribution, or some other family of distributions, is more appropriate for describing floods. This, however, would represent a challenge for which there is little guidance in the scientific literature. Furthermore, it is not clear that much would be gained by incorporating that uncertainty, as compared to working with the LP3 distribution and exploring thoroughly the joint uncertainty of all three parameters of the distribution. In the committee's judgment, use of the LP3 distribution is reasonable, but a procedure should be developed that more adequately captures the true posterior distribution of the LP3 distribution parameters in a Bayesian sense (Bobee and Ashkar, 1991; Chowdhury and Stedinger, 1991; Stedinger, 1983a).

Neglecting Skew Uncertainty

Hydrologic uncertainty is simpler to deal with than other sources of uncertainty when the analysis is based upon a stationary gauged record. For the most part, hydrologic uncertainty in estimators of the parameters is determined by the limited length of the flood series available to estimate the values of the parameters of the LP3 distribution commonly used in the Corps's method. In that sense, the uncertainty is objective and is described by standard statistical sampling theory (Chow et al., 1988; Chowdhury and Stedinger, 1991; IACWD, 1981;).

When a flood record must be corrected for land use changes, storage, or channel changes, the length of record is still likely to be the primary determinant of hydrologic uncertainty, although subjective assessments of the quality of any adjustments to measured flows are also important. Possible nonstationarity due to subtle shifts in climate and storm paths, which are difficult to document, is sometimes a concern. If regional relationships are used to develop flood curves, then the corresponding estimates of prediction error should be employed (Tasker and Stedinger, 1989).

In its analysis for gauged sites, the Corps bases its description of hydrologic uncertainty upon the confidence interval calculation procedure in *Bulletin 17B* (IACWD, 1981), a document that contains procedures that federal agencies agreed to employ in the mid-1970s. The *Bulletin 17B* procedure for calculating confidence intervals employs the assumption that the coefficient of skewness of the logarithms of the floods is correctly specified, independent of the data (Stedinger, 1983b). The actual coefficient of skewness employed is generally a weighted average of the at-site sample skewness and a regional or generalized skewness estimator (IACWD, 1981). The weighted skewness estimators clearly incorporate estimation error because of sampling error in the at-site skewness estimators and also in the regional skewness estimators (McCuen, 1979; Tasker and Stedinger, 1986). As a result, the intervals calculated with the *Bulletin 17B* procedure are too small. Equations that incorporate variability in weighted skewness estimators are available (Chowdhury and Stedinger, 1991; Stedinger et al., 1993) and should be used.

It is not appropriate for the Corps to ignore the large uncertainty in the estimated coefficient of skewness of the LP3 distribution used to describe flood risk. Clearly, an operational procedure to adjust uncertainty in the skewness coefficient can be developed using available sampling theory (Bobee and Ashkar, 1991).

Errors in Flood Frequency Curves
Derived from Rainfall–Runoff Modeling

Although classical flood–frequency analysis is based on analysis of observed discharge data, practical requirements in Corps flood damage reduction project planning often lead to the use of rainfall–runoff modeling to synthesize the flood–frequency curves used in planning studies. In its risk analysis, the Corps treats these synthetic flood–frequency curves as being of equivalent accuracy to flood–frequency curves derived from a graphical fit to a set of observed flood discharge data. However, synthetic flood–frequency curves have additional error in them because of uncertainties involved in rainfall–runoff modeling. This additional error can be estimated by the difference between the flood–frequency curve developed from observed discharge data at a gauging station, and a flood–frequency curve synthesized at the same location by rainfall–runoff modeling. The committee recommends that the Corps further examine this approach as a way of quantifying the additional error introduced by using rainfall–runoff modeling to produce flood frequency curves at ungauged sites.

Errors in the Stage-Discharge Relationship

The Corps's Engineering Manual (EM) for risk analysis, EM 1110-2-1619 (USACE, 1996b), describes several methods for estimating errors in the stage–discharge relationship: using variability of observed gaging data and rating curves at stream gaging stations, comparing observed high-water marks during historical floods with those produced by water surface profile computation, by sensitivity studies of the effects of variations in Manning's n on water surface elevation, and by standard estimates of error in cross-section profile elevations. When these alternative methods are applied, there can be substantial differences in the results. The Corps's Engineering Manual recommends taking the highest and lowest estimated water surface elevation for a specified discharge from all the methods and dividing the result by 4 to estimate the standard deviation, on the basis that two standard deviations above and below the mean contain 95 percent of the variability of normal distribution of errors.

GEOTECHNICAL RELIABILITY

Geotechnical reliability is an important consideration in flood damage reduction projects involving levees. The reliability computation accounts for the potential of the levee to breach through soil failure even when the water surface elevation is not sufficiently high to overtop the levee. The method for quantifying geotechnical reliability is described in Chapter 4 and is applied in the case of the Grand Forks case study described in Chapter 5.

As is the case with other parts of the Corps's risk analysis, the geotechnical reliability model does not separate natural variability from knowledge uncertainty. This has important implications. The natural variability (treated as stochastic variability) arises from spatial variations in soil conditions and levee construction. The knowledge uncertainty arises from modeling assumptions made in calculating geotechnical performance. The former varies independently, or nearly so, from one reach to another. The latter is systematic, or nearly so, across all reaches. Probabilistically, the natural variability in site conditions may be independent from one reach to another (presuming large enough reaches to neutralize spatial autocorrelation effects in modeling natural variations), while the knowledge uncertainties are highly correlated. The important implication arises when one calculates the probability of at least one levee failing anywhere along the river. If the reaches are independent, then conceptually, this probability rises according to a relation of the form,

$$\Pr\{\geq 1 \text{ levee failure}\} = 1 - (1 - p)^n, \qquad (6.1)$$

where p is the probability of failure in any one reach and n is the number of reaches. This probability rises quickly with increasing n. Alternatively, if the reaches are perfectly correlated, the probability of at least one levee failing anywhere along the river is given by Pr {≥1 levee failure}= p. Just as natural variability and knowledge uncertainty are distinguished in the hydrologic and hydraulic modeling, so too should these types of uncertainty be treated distinctly and differently in the geotechnical modeling.

Levee failures caused by seepage under or through the levee—and even slope instabilities, which are influenced by pore pressures internal to the levee—depend not just on water height, but also on the duration of flooding. Flood duration is not considered in the current geotechnical reliability model, and indeed may be difficult to accommodate in the risk analysis method at all. To this extent, the geotechnical reliability model is an approximation. This is not a crucial limitation, given other approximations in the risk method, but the geotechnical calculations might be significantly improved if duration information were generated by the hydrology and hydraulics model and incorporated in the geotechnical model.

Although the geotechnical reliability model is a sound first step, it is also a new approach. The model would benefit greatly from field validation. The nation has many years of experience with levee performance and, unfortunately, also with levee failures. Much of this experience is documented, and much is accessible to federal agencies. The Corps should undertake statistical ex post studies to compare predictions of geotechnical failure probabilities made by the reliability model against frequencies of actual levee failures during floods.

It should also be pointed out that geotechnical failure of a levee(s) is not the only way in which a flood damage reduction project might fail; for example, hydraulic facilities may fail or detention basins may overflow. The committee thus recommends that the Corps also conduct statistical ex post studies with respect to the performance of other flood damage reduction measures (e.g., detention basins, hydraulic facilities). These studies should be conducted to identify the vulnerabilities (failure modes) of these systems and to verify engineering reliability models.

ECONOMIC PERFORMANCE

The assessment of economic performance of a project is measured by project net benefits. Net benefits are the difference between benefits and costs, where benefits are defined as the reduction in flood damage resulting from the project. Assessment of economic performance builds upon hydrologic, hydraulic, and geotechnical factors that enter into the assessment of engineering performance, plus the computation of flood damage to structures or other activity in the floodplain. While engineering performance is focused on risk at each damage reach, economic assessment is more complex, involving the integration of information at several spatial scales.

In the Beargrass Creek case study (Chapter 5) there are five spatial scales of analysis, as shown in Table 6.2. The three main scales are the following:

- project scale at which all the economic analysis is summarized,
- damage reach scale used for most analysis in HEC-FDA, and
- structure scale where the assessment of damage to structures is carried out.

The single project is subdivided into 21 damage reaches, which contain 2051 structures. In between these three main scales are two others:

- three main river reaches, used to identify the components of the project plan, and
- hydraulic cross sections (263), used for determining the water surface profile.

At Beargrass Creek, there are, on average, about 100 structures per damage reach.

Monte Carlo simulation is repeated, independently, 100 times for each structure. Four variables are randomized for each structure: first-floor elevation, value of the structure, value of the contents, and other valuesof the facility. The results of these simulations are aggregated by damage category (e.g., single-family residential, multifamily residential)

TABLE 6.2 Scales of Spatial Analysis in the Beargrass Creek Study

Spatial Scale	Uses of Spatial Unit
Project (1)	Expected Annual Damage (EAD) and Benefit-Cost analysis
Main River Reaches (3)	Incremental analysis to get National Economic Development (NED) plan
Damage Reaches (21)	Basic unit for analysis using HEC-FDA, including flood–frequency curves
Hydraulic Cross Sections (263)	Water surface elevation profile computation
Structures (2,051)	Structure inventory

to form a damage–stage function with uncertainty for each category of damage. The category damage functions are aggregated to form a damage–stage function for each reach. In effect, about 400 random variables are accumulated into a single random variable for each reach, which is damage as a function of stage. This process is illustrated in Figure 6.2 for damage to multifamily structures in damage reach SF-9 at Beargrass Creek.

Development of the distribution of damages considering each structure in a reach is the first stage of the risk analysis. In the second stage, Monte Carlo simulation is done 10,000–100,000 times for each damage reach. Three functions are randomized in this second stage: the flood discharge–probability curve for flood magnitude, the stage–discharge curve for water surface elevation, and the damage–stage curve for the degree of damage. For each randomization in stage 2, the damage curve is integrated across the probability range of flood severity to determine the expected annual damage for that replicate. The ensemble of all the expected annual damage estimates so formed from the 10,000–100,000 simulations yields the probability distribution of the flood damage at that reach for each plan being considered.

Once this analysis is complete, the expected annual damage (EAD) for the project is found by summing the mean values of the expected annual damages for each reach. Statistics of the percentile distribution (25%, 50%, 75%) of the expected project damage for this plan are currently computed by summing the corresponding percentile values of the distributions of each reach. Hence, the values of 22 random variables at the damage reach scale are aggregated to form a single random variable at the project scale, which defines the flood damage for a given plan.

The Engineering Regulation that guides the evaluation process for flood damage reduction projects (USACE, 1996b) calls for a table showing the probability distribution of the reduction in expected annual damage due to the project plan, with percentiles at 5%, 25%, 50%, 75%, 95%, and it calls for a similar table for the percentile distribution of project plan net benefits. The reduction in expected annual damage is found in each case by subtracting the damage with the plan from that without the plan. Thus, a percentile distribution describing project benefits is computed as the difference between with-plan and without-plan percentile distributions of flood damage.

The procedure of adding and subtracting percentile values of statistical distributions of flood damage as if they were arithmetic quantities is statistically unsound because it ignores the degree of interdependence or correlation among these distributions.

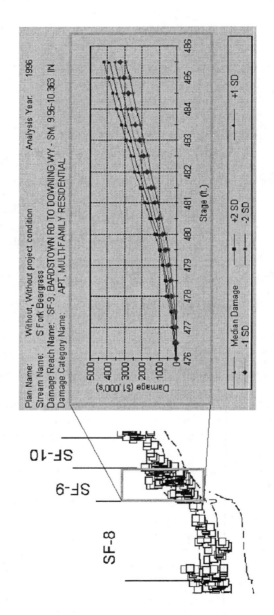

FIGURE 6.2 Aggregation of damage to multifamily structures in Reach SF-9 at Beargrass Creek.

INTERDEPENDENCE IN RISK ANALYSIS
FOR FLOOD DAMAGE ASSESSMENT

The principal variables of the flood damage assessment—flood discharge, stage, and damage—are functionally dependent on one another through the stage–discharge curve and the damage–stage curve at each damage reach. Because all these curves also contain uncertainty determined independently for discharge, stage, and damage, the end result is that discharge, stage, and damage are strongly correlated variables. The principle of transforming the flood–frequency curve into a stage–frequency curve, and then into a damage–frequency curve takes account of that interdependence.

However, care must still be taken to sample each function correctly, and there are several choices as to how the derived damage functions should be selected. The choice will affect the computed variance of expected damages for the reach due to economic uncertainty. Understanding the correct error structure for generating damage functions requires understanding correlation between errors in structural value and content value at individual structures, and it requires understanding correlation errors in first-floor elevations of structures at different locations. These issues are not addressed in the current implementation of HEC-FDA.

It appears that apart from treating the interdependence and the interaction among the three main variables at each damage reach, the current version of the risk analysis procedure assumes that all the random variables are statistically independent. In the Engineering Manual describing the procedure (USACE, 1996a), correlation is mentioned only twice: once on p. 5-1 ("Any correlation of separate factors should also be considered in the analysis and accounted for in the combination of individual uncertainties") and once on p. 5-4, which shows a relationship between the uncertainty in the stage–discharge relationship and the slope of the river. The flow charts describing the computational procedure in this manual make no further reference to correlation or interdependence, and the input data to the HEC-FDA program do not include correlation coefficients or other representations of interdependence.

It would appear, at least on the surface, that some of the input variables used in the analysis are highly dependent. For example, the value of a structure and the value of its contents would usually be correlated, such that if a particular structure is actually more valuable than an estimate would indicate, most likely its contents are also more valuable than the estimate. Similarly, if the first-floor elevation of a structure is in error by some margin, say one foot too high, it is likely the first-floor elevation of an adjacent structure is also too high by a similar amount. This is because both structures have their elevations estimated from topographic mapping, and errors in mapping tend to be systematic over some distance.

These two examples introduce two different types of interdependence—cross-correlation between two variables (e.g., value of structure and value of contents) and spatial interdependence of the same variable at two locations (e.g., first floor elevations of adjacent structures). The spatial correlation issue has special significance in flood damage assessment because of the different spatial scales at which the various components of the problem are being analyzed (Table 6.2).

CORRELATION LENGTH

There are several measures to describe the degree of spatial correlation ρ_w as a function of distance w between two locations. The simplest of these is the correlation length, L_r, also called the scale of fluctuation (Vanmarcke, 1983), which is the integral of the area under the spatial correlation function (Figure 6.3) and which is calculated as,

$$L_r = \int_0^\infty \rho_w dw. \tag{6.2}$$

Because correlation is a dimensionless quantity while distance has the dimension of length, the value of L_r also has the dimensions of length; hence the name correlation length. There are other measures of spatial correlation available, most notably the variogram used in geostatistics (Cressie, 1993), but correlation length sufficiently describes the degree of spatial correlation for purposes of this discussion.

In the Beargrass Creek study, for example, the first-floor elevations of the structures were estimated from 1-inch-to-100-feet topographic mapping, using a contour interval of 2 feet. As a check, the first-floor elevations of 195 structures were determined by land surveying from which it was concluded that "the average of the absolute values of the differences between the estimated and surveyed first floor elevations for this sample is 0.62 ft."(USACE, 1997c, p. B-5). Because the (x,y) location of each structure is determined through surveying, as well as the elevation z, it follows that the horizontal distance between each pair of structures can be found, and a spatial correlation function of the errors

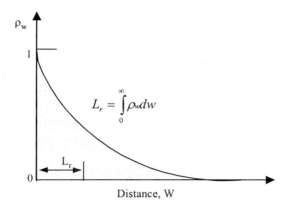

FIGURE 6.3 The spatial correlation function, ρ_w, and correlation length, L_r.

could have been constructed with the information gathered in the flood damage reduction study.

SPATIAL AGGREGATION

Spatial aggregation refers to the assembly of a set of random variables within a defined region into a single variable representative of that region. This happens within each damage reach when damage to structures is aggregated into a damage–stage curve for the reach, and it happens at the project scale when statistical measures of damage at the reach level are accumulated to form measures of the damage for the project.

The three basic spatial scales of analysis in flood damage assessment are shown in Figure 6.4 for the South Fork of Beargrass Creek. This figure shows water surface profiles for four different flood severities, with the locations of the structures superimposed. The river reach of approximately 12 miles in length is divided into 15 damage reaches for statistical analysis, averaging 0.8 miles in length. If a larger-than-anticipated precipitation depth corresponds to a specified exceedance probability, it will result from an unusually severe storm over most, if not all, of the drainage area. The corresponding uncertainties in the project flood discharge are those likely to systematically propagate throughout all the damage reaches.

Similarly, the laws of hydraulics impose a significant degree of continuity of water surface profiles through damage reaches. The backwater effect of the bridge constriction at the upper end of damage reach SF-1 propagates some distance upstream, perhaps through damage reaches SF-2 and SF-3 but probably not any farther. Backwater effects extend for longer distances in flatter terrain. Although there are likely to be correlations between errors in damages between adjacent structures within a damage reach, it is unlikely that these correlations in structure damage would extend over long distances. It is expected that the correlation length for errors in the flood discharge estimates may be on the order of the length of the whole river reach, and that the correlation length for errors in flood stage is of the order of one damage reach. The correlation length of errors in structure damage, however, is probably much less than the length of a damage reach.

An important principle in spatial aggregation is that when quantities are averaged over nonoverlapping intervals, the correlation between the averaged quantities goes to zero when the averaging interval becomes much larger than the correlation length (Vanmarcke, 1983, p. 199). In other words, if the length of the damage reach, ΔL, is significantly larger than the correlation length of the errors in structure damage, L_r, then the stage–damage curves in adjacent damage reaches can be considered statistically independent. However, the representation of the variance of variables within each reach will still be in error.

Because the correlation length of errors in hydrology and hydraulics is probably much longer than that of errors in the damages, it is likely that the assumption of independence of the flood–frequency curves and stage–discharge curves for each damage reach is invalid. Indeed, the opposite argument can probably be made, namely that errors in flood–discharge and stage–discharge relationships are highly correlated between adjacent damage reaches.

In the HEC-FDA program (USACE, 1998a, p. 7-15) the 25 percent, 50 percent, and 75 percent values of the damage reduced by project plans are displayed, computed by subtracting the corresponding percentiles of the damages with and without the plan and summing the results over all reaches in the project. This process of adding and subtracting percentiles of distributions as if they were expected values is not statistically sound. The values of percentiles of distributions depend on the variance and

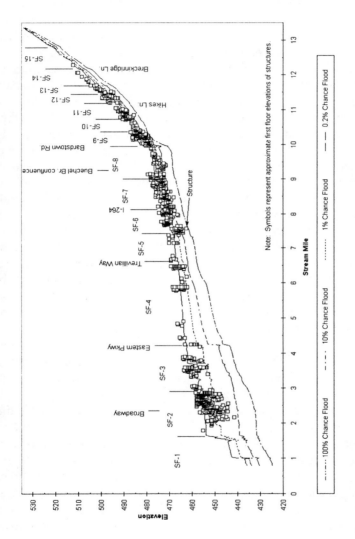

FIGURE 6.4 Spatial scales of analysis in flood damage reduction studies for the South Fork of the Beargrass Creek.

other moments of the distribution. As shown in Equations 6 and 9 of Appendix D, the variance of a sum or difference of a pair of random variables depends in part on the correlation between them. Because these correlations have not been accounted for in the risk analysis procedure, there is no guarantee that the percentile variabilities of project flood damages and benefits determined in the current manner correctly reflect the actual variability of the damage reduced by the project plan.

The procedure of adding up the percentile values at each reach to get values for the project is valid in the event that the damage distributions at each reach are perfectly correlated. In that case, if the project damage is at its 25 percent value, the reach damages are also all identically at their 25 percent values, and the values for the reaches can legitimately be summed to give the value for the project. The long correlation length of the hydrologic and hydraulic uncertainties makes it likely that the reach damages are correlated, but that correlation will not be perfect. Indeed, the Monte Carlo simulations for each reach are based on the assumption that the reaches are independent.

There is thus an internal contradiction in the method—to determine variability in project reduced damage and net benefits, the summation made over values at the damage reaches relies on the assumption that reach damages are perfectly correlated from reach to reach; however, the computational procedure used to create the damage values treats each reach completely independently of its neighboring reaches upstream and downstream. If the reaches are perfectly correlated, they all behave in complete statistical cohesion, with damage errors everywhere being equally higher or lower than expected values. Such perfect correlation is rare in nature and unlikely in this instance because of the random events that occur as a flood moves through a river system.

Consider again the basic questions asked at the beginning of this chapter. For risk analysis of economic performance, is the computational procedure theoretically valid? The answer is that it is not. The basic procedure of integrating the discharge, stage, and damage through Monte Carlo simulation is reasonable; however, there are several unresolved issues regarding correlation and spatial aggregation at the structure scale, correlation from the structure scale to the damage reach scale, and correlation from the damage reach scale to the project scale. The second question raised at the beginning of the chapter is whether there are ways of improving the precision of the probability estimates. The following section suggests some ways of doing that.

COMPUTATIONAL ALTERNATIVES
TO MINIMIZE CORRELATION EFFECTS

The addition of formal measures of risk to the flood damage assessment procedure is a necessary step, which recognizes the inherent uncertainty in the computations and their results. It is also a complex procedure. Unfortunately, the rigors of treatment of combinations of random variables allow only a limited number of standard operations, far fewer than if the uncertainty in the variables is ignored.

In effect, the current risk analysis procedure evolved from an earlier deterministic procedure by randomizing uncertain variables, solving equations for each randomization, and then averaging over the results to get expected values. This is a valid process provided that correlation is introduced where appropriate in generating random variables. There may also be some ways to change the computational process to minimize or eliminate the effects of correlation, as discussed below.

Determine the Scale of Randomization

The Monte Carlo analysis should be conducted at the spatial scale at which the results are required. If results are required for the project, the Monte Carlo replicates should be constructed at the project. In other words, a Monte Carlo *project realization* would be defined by a set of random variables specifying everywhere in the project the errors in flood discharge, stage, and damage to each structure. Then, the resulting engineering and economic performance measures can be determined for this realization. By repeatedly generating such project realizations, the statistical variability in the project engineering and economic performance measures can be determined. As a part of this project realization calculation, corresponding performance measures can also be determined for each damage reach so that the distribution of damages by category and reach could be defined. In addition, this approach also allows quantification of the project benefits and of the reduced risk of flooding at each individual structure in the project. Advances in computing power mean that methodological approximations employed in past practice, such as aggregation of all the structure computations to the damage reach scale, may no longer be needed.

Introduce Correlation in Monte Carlo Simulation

Correlation in random variables should be introduced where necessary by selecting the random variables in Monte Carlo analysis using a
system for generating correlated sets of variables rather than using a
system that generates independent sets of variables. This practice is
common in groundwater modeling where correlated sets of hydraulic
conductivities and other aquifer properties are generated by Monte Carlo
simulation. Similarly, there is a long history of generating correlated
random variables to describe stream flow in a basin or region (Salas,
1993).

Randomize Structures Jointly

The damage reach mainly functions as a place of aggregation of
structures to an index location, and this should be considered in the context of other structures. If the view shown in Figure 6.2 is adopted, each
structure can be left at its original location and can be attributed with the
computed water surface elevations with and without the plan for different
flood severities. In that manner, each structure can be randomized in
association with its near neighbors. The damage reach is really a label-
ing device to show how damage varies along the river. If the Corps
elects to retain the two-stage Monte Carlo simulation process (simulate
the structures first, then the damage reaches), the method should generate
appropriate flood stage–damage functions for each reach. If a set of
structure elevations, structure values, and content values are generated
jointly to create a realization of the economics for a reach, one obtains
one realization of the damage function for the reach. These empirical
reach–damage functions computed at the first stage can be generated and
stored for use at the second-stage reach-level Monte Carlo analysis.
Thus, instead of using some artificial method for generating reach–damage functions, one can sample from the 100 functions generated randomly at the first stage. These would be combined with descriptions of
flood–frequency and flow–stage relationships. This would retain the
economic spatial interdependencies within a reach for use in the second-
stage analysis.

Randomize Hydrology and Hydraulics for River Reaches

The flood hydrology and hydraulics should be randomized at the scale of the river reach rather than at the damage reach. In this manner the flood profiles for a given event would move randomly up and down in cohesion over the whole reach, as regional hydrology and hydraulics suggest they should. This concept would also allow quantification of uncertainty in the spatial extent of the floodplain boundary. There is some upper limit on how long a river reach can be to be considered as a single statistical entity. The whole Upper Mississippi River basin, for example, would need to be subdivided significantly for such an analysis.

Analyze Statistical Variability in Project Benefits Rather than Damage

For the economic analysis, the project benefits are measured by the reduced damage at each structure. For each project plan and cycle of randomization, the issue of prime concern is the difference between the damage with the plan compared to the damage without it. That calculation is presently conducted by determining the flood damage for all structures in a damage reach and then aggregating over the damage reaches to get the expected annual damage for the project. Finally, after all aggregation is complete, the project benefits are determined by taking the difference in damage with and without the project plan. An alternative to this computation is to consider the project benefits structure by structure and then to aggregate those benefits over the project.

Figure 6.5 shows the water surface profiles for existing conditions and under a proposed flood damage reduction plan for the 10-year and 100-year floods in a damage reach. The box symbols shown in the figure are the first-floor elevations of the structures in this damage reach. In each case, the proposed water surface profiles (shown by the solid lines) lie beneath the existing profiles (shown by the dashed lines), more so at the downstream end of the reach than at the upstream end. The difference in elevation between the water surface profile and the first floor elevation of the structure measures the depth of flooding at each structure, which is then related to the flood damage.

If the flood severity is indexed by its annual probability p ($p = 1/T$, where T is return period of the flood) and by its location x in river miles along the river, then the damage at that location can be denoted as $D(p, x)$. If the damages under existing conditions and with-plan conditions are

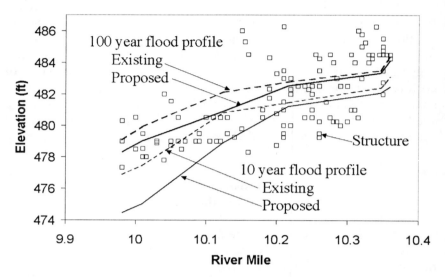

FIGURE 6.5 Comparison of water surface profiles for existing conditions and with a proposed flood damage reduction plan. This profile is from damage reach SF-9, Beargrass Creek.

denoted by D_e, and D_w, respectively, then the project benefit B for a particular flood at location x is given by

$$B(p,x) = D_e(p,x) - D_w(p,x). \qquad (6.3)$$

The annual benefits of the plan for this severity over the whole project are then determined by integration over the length L of river affected by the project:

$$B(p) = \int_0^L B(p,x)dx. \qquad (6.4)$$

If the project benefit is determined for a set of floods, p, then the expected annual benefit (EAB) of the project considering all floods is found by integrating the benefits over the range of p:

$$EAB = \int_0^1 B(p)\,dp \qquad\qquad (6.5)$$

By comparing this value to the project costs, the economic value of the project is determined.

This method of determining the project benefits does not require any more information than is used in the current procedure. It just manipulates that information according to a different computational sequence. This method has some advantages:

1.   It permits localized benefits and the risk of flooding to be determined down to the level of individual structures, instead of lumping all structures into a damage reach value

2.   It visually shows the difference between the existing and proposed flood profiles so that the effects of the plan at particular locations and for different flood severities can easily be seen.

3.   The computation in Equations 1–3 in Appendix D can be done for each Monte Carlo replicate of the flood water surface profile considering random errors in the profile and in the elevations of the structures. By summing across the whole project the damage reduced at each structure, Monte Carlo replicates for benefits for the whole project can be found, and their mean and percentile distribution can be determined once the Monte Carlo simulations are complete. This is a valid statistical procedure not confounded by the problem of aggregating probability distributions from each damage reach as is now done in HEC-FDA.

4.   By calculating the difference between the existing-conditions and with-plan damage estimates at each structure, the effects of errors in elevation and in values of the structure and its contents at a particular structure will be diminished. In other words, if the first-floor elevation of a structure is in error, the damage to a structure under existing and proposed conditions will be affected by nearly the same amount, so when the benefits are determined by taking the difference in the two damage values, the effect of the error in first-floor elevation will nearly be cancelled out. Similar considerations apply to the effects of all other errors being considered. It is also possible that this approach will significantly diminish the effects of correlation in the errors.

It is possible that this approach to determining the variability in the project benefits will result in significantly smaller variance estimates than those currently being determined because the current estimates are being made by taking the differences between two large and highly variable numbers—the project damage estimates with and without the plan after all randomization is complete. The key point is that what matters in economic assessment of flood damage is not so much the absolute magnitude of the flood damage, but rather the extent to which the proposed plan will reduce that damage. The committee recommends that the computational procedure for risk analysis of economic performance focus on the uncertainty in the project benefits rather than on the uncertainty in the project flood damage.

Statistically Compare Net Benefits from Alternative Plans

If an accurate estimate of the variance of the damage reduced by a project plan can be found, it could be used to determine whether one project plan has expected annual net benefits that are better than those of another plan in a statistical sense. In other words, it is possible that the addition of a marginal component to a plan slightly increases the expected annual net benefits but that the increase is not statistically significant when compared to the uncertainty of estimating the annual benefits. In this manner, the statistical measures of the variability of the benefits could have a greater influence on project decision making than they currently do.

It is perhaps impossible to achieve complete statistical rigor in a complex problem like risk analysis for flood damage assessment. The key is to be able to construct an analysis capturing the critical aspects of the statistical variability of the problem without excessive investment in details, which may have an insignificant impact on the final risk estimates. Alternative computational methods need to be tested in further case studies of flood damage reduction projects so that a deeper appreciation is obtained of the advantages and limitations of these alternatives in risk analysis.

7

Levee Certification

The National Flood Insurance Act of 1968 sought to reduce suffering and economic damage from floods. This congressional act created the National Flood Insurance Program (NFIP). Revisions to this act have included the Flood Disaster Protection Act of 1973 and the Flood Insurance Reform Act of 1994. The National Flood Insurance Program was originally placed under the authority of the secretary of Housing and Urban Development (program authority today lies with the administrator of the Federal Insurance Administration). In 1979 the Federal Insurance Administration and its programs, including the National Flood Insurance Program, were transferred to the newly created Federal Emergency Management Agency (FEMA).

HISTORY OF LEVEE CERTIFICATION

Congress sought to accomplish two main goals in the National Flood Insurance Act of 1968 and the Flood Disaster Protection Act of 1973. Congress wanted property owners to purchase flood insurance to (1) provide them with financial relief should they suffer losses in a flood and (2) lessen the financial burden on federal, state, and local governments to provide grants and low-interest loans to cover the losses of uninsured property owners. These acts also sought to reduce damage from moderate-sized floods by encouraging construction of levees and other flood damage reduction structures. To achieve these goals, Congress offered incentives to communities whose flood damage reduction structures were certified to prevent damage from a moderate-sized flood. Levee certification allows communities and structures to be removed from "Special Flood Hazard Areas" (defined as areas subject to inundation during a 100-year flood), thereby removing mandatory flood insurance purchase regulations (the 1973 Flood Disaster Protection Act requires those who are buying, building, or improving property in special flood hazard areas within NFIP communities to purchase flood insurance as a prerequisite for federal financial assistance (e.g., loan, grant, disaster assistance) when the building or personal property is the subject of or security for such assistance). Levee certification could thus exempt a community from flood insurance purchase requirements and could also remove some local land use restrictions. As certification could exempt a community from thousands, perhaps millions, of dollars of flood insurance premiums, this certification procedure (and the risk analysis therein) has great local economic and public policy significance. Unfortunately, Congress gave only vague guidance as to what size event the levees were to withstand.[1] The National Flood Insurance Act of 1968, as amended, and the Flood Disaster Protection Act of 1973, as amended, do not contain detailed statements regarding the hydrological or statistical significance of the 100-year flood, and contain but one significant reference to the 100-year flood: "Notwithstanding any other provision of law, any community that has made adequate progress, acceptable to the Director, on the construction of a flood protection system which will afford flood protection for the one-hundred year frequency flood as determined by the Director, shall be eligible for flood insurance under this chapter . . . . at premium rates not exceeding those which would be applicable under this section if such flood protection system had been completed" (42 U.S.C. 4001 *et. seq*; the "Director" is the Director of FEMA).

Congress was less than explicit in providing direction to the Corps concerning the desired level of flood protection for the nation. Congress directed the Corps to map 100-year floodplains as the areas that needed protection (Box 7.1 describes the adoption of the 100-year flood for

---

[1] For levees for which risk analysis has not been performed, FEMA continues to specify (as of May 2000) that a minimum of 3 feet above the base flood is needed for certification (44 CFR 1 §65.10). Some requirements related to additional freeboard (such as an additional 1 foot of freeboard near structures and ½ foot of freeboard at the upstream end of the levee) and requirements related to maintenance, closures, embankments, foundations, and drainage are also imposed.

floodplain management and insurance purposes). Congress could have specified the 200- or 500-year floodplains. The committee infers that Congress desired protection against the 100-year flood. If they had wanted protection against the 200- or 500-year floods, they would have directed the Corps to map the 200- or 500-year floodplains.

Confusion crept in, however, concerning whether levees were to protect against the 100-year flood or whether they were to lower the likelihood of flooding to 1/100 per year. FEMA focused on protecting against the 100-year flood and (for certification purposes) required that 3 feet of freeboard be added to the stage (height) of the 100-year flood, to be confident that the levees could pass this flow. In the absence of risk analysis, freeboard was sensible: levees built only to the elevation of the 100-year flood would not necessarily survive it (because of waves, wind, and other uncertainties). Adding 3 feet of freeboard to levees, however, turns out to be equivalent (on average) to reducing the likelihood of flooding to roughly 1/230 per year, a much more stringent standard than Congress apparently intended.

The Corps's use of risk analysis resulted in a qualitative change in both the theory and practice of certifying levee systems. Risk analysis permits the Corps and FEMA to address problems resulting from uneven safety levels and from expenses inherent in the criterion of having the levee sized to the 100-year flood plus 3 feet of freeboard. As shown in Column 2 of Table 7.1, the 3-feet-of-freeboard criterion resulted in very different levels of protection in different communities. For example, a levee satisfying this criterion would have only a 45.3 percent chance of passing a 100-year flood in East Peoria, Illinois, but would have a 99.9 percent chance of passing a 100-year flood in West Sacramento, California; Portage, Wisconsin; and Hamburg, Iowa. When the Corps and FEMA used 3 feet of freeboard to certify a levee, they were using a criterion that provided less protection in some communities than in others. Similarly, the criterion resulted in excessive expenditures on some levees.

Higher levees may provide greater levels of flood protection, but levee heights are limited by the additional costs of higher levees and by their ecological and aesthetic effects on water bodies and floodplains. In particular, the three communities with more than a 99.9 percent chance of passing a 100-year flood (see Table 7.1) have flood damage reduction systems that, all other things being equal, are too expensive and impose high ecological and aesthetic costs. Communities whose levees have only a 45.3 percent chance of passing the 100-year flood have a much

BOX 7.1
Why the 100-year Flood?
Gilbert White Illuminates

The concept of the 100-year flood is central to the National Flood Insurance Program and to many of the Corps's flood damage reduction activities. Hundreds of government officials administer and work within these flood mitigation and damage reduction programs, to which millions of taxpayer dollars have been devoted. Many consultants are employed in mapping the nation's 100-year floodplains and scores of university professors analyze the hydrological, statistical, and public policy implications of the 100-year flood. Given the economic and social importance of these efforts, one would assume that the selection of the 100-year flood as a defining hydrological event is based on sound scientific and statistical foundations.

Gilbert White, professor emeritus of geography at the University of Colorado, is widely recognized as a leader in promoting sound U.S. flood management strategies. In 1993, Professor White provided an oral interview to Martin Reuss, the Corps of Engineers's senior historian. In that interview, White's response to a question about the selection of the 100-year flood sheds some light on the rationale for its selection. Given his knowledge of and experience in U.S. floodplain management, Gilbert White's account may be among the better explanations we have for the prominence of the 100-year flood in U.S. floodplain management and policy.

In response to the question, "How do you take into account the so-called catastrophic flood—the once-in-100-years flood?", White stated:

"There was a very interesting development of the notion that there could be a flood of sufficiently low frequency that no effort should be made to cope with it. The Federal Insurance Administration picked one percent [or] a recurrence interval of a hundred years. And some of us were involved in that because we recognized they initially had to have some figure to use. The one-percent flood was chosen. I think Jim Goddard and TVA colleagues would be considered parties to the crime. With the lack of any other figure, the concept taken from TVA's "intermediate regional flood" seemed a moderately reasonable figure. We generally use the term "catastrophic flood" for events of much lesser frequency.

This goes back to my earlier criticism of the FIA and its determination to cover the country promptly. In covering the country promptly they established one criterion—the 100-year flood. I think it would have been much more satisfactory if they had not tried to impose a single criterion but had recognized that there could be different criteria for different situations. This could have been practicable administra-

---

BOX 7.1 Continued

tively even though a federal administrator would say it's far easier, cleaner, to have a single criterion that blankets the country as a whole.

What's the effect of having a criterion of 100 if in doing so a local community is encouraged to regulate any development up to that line and then to say we don't care what happens above that line? We know that in a community like Rapid City the floods were of a lesser frequency than 100 years, and a community ought to be aware of this possibility.

A simplified national policy tended to discourage communities from looking at the flood problem in a community-wide context, considering the whole range of possible floods that would occur.

So I would say that any community ought to be sensitive to the possibility of there being a 500-year flood or 1,000-year flood. It should try to consider what it would do in that circumstance, and wherein it could organize its development so that if and when that great event does occur it will have the minimum kind of dislocation."

Gilbert White referred to several risk-related topics addressed in this report. For example, his comment regarding the value of using different criteria for different situations buttresses the Corps's adoption of risk analysis techniques and the abandonment of the levee free-board principle. As White pointed out, different geographical areas are subject to different levels of flood risk and uncertainty and thereby require different margins of safety. The committee also agrees with Professor White's comments regarding flood hazard preparedness for floods of *all* magnitudes. This committee recommends that rather than focusing on a single event—the 100-year flood—that the Corps examine the risks of flooding from the full range of possible floods.

---

lower degree of flood protection. A community forced to upgrade its levees to the 99.9 percent level would likely find levee certification very expensive. A community that had its levees certified despite having only a 45.3 percent chance of passing the 100-year flood would be lured into complacency, as they are likely to experience a costly flood resulting in losses to individuals and in losses to the NFIP.

An additional difficulty with the 3-feet-of-freeboard certification criterion is that it could impose unnecessarily large costs on communities. Congress instructed the Corps to build water projects only if the benefits exceed the costs. In its flood damage reduction studies, the

TABLE 7.1 Elevation, Freeboard, and Expected Level of Protection Provided by Various Methods of Levee Sizing[a]

Levee Project	Base Elevation (ft)	FEMA 3 ft-Freeboard Conditional Nonexceedance Probability	NED PLAN[b]			1% Expected AEP[c]		
			Elev (ft)	Freebd (ft)	Return period[d] (yrs)	Elev (ft)	Freebd (ft)	Return period[d] (yrs)
Column	(1)	(2)	(3)	(4)	(5)	(6)	(7)	(8)
Pearl River, MS	41.6	0.976	47.0	5.4	770	41.8	0.2	100
American River, CA	46.1	0.919	52.0	5.9	230	47.1	1.0	100
West Sacramento, CA	29.2	0.999	33.5	4.3	1,670	29.6	0.4	100
Portage, WI	795.3	0.999	797.0	1.7	10,000	795.6	0.3	100
Grand Forks, ND	831.4	0.908	N/A	N/A	N/A	831.5	0.1	100
Hamburg, IA	909.2	0.999	911.5	2.3	910	909.8	0.6	100
Pender, NE	1,326.3	0.763	1,330.0	3.7	380	1,328.0	1.5	100
Muscatine, IA	557.8	0.901	561.5	3.7	330	558.8	1.0	100
Cedar Falls, IA	861.7	0.900	866.0	4.3	360	862.6	0.9	100
Sny Island LDD, IL	471.1	0.567	N/A	N/A	N/A	471.5	0.4	100
East Peoria, IL	455.1	0.453	462.6	7.5	10,000	458.3	3.2	100
Guadalupe River, TX	54.9	0.872	56.8	1.9	110	56.5	1.6	100
White River, IN	712	0.980	713.2	1.2	250	712.3	0.3	100
Mean		0.864		3.8	380		0.9	100
Median		0.908		3.7	380		0.6	100
Minimum		0.453		1.2	110		0.1	100
Maximum		0.999		7.5	1,000		3.2	100

[a]Bold faced type indicates levee height required for certification within the NFIP.
[b]NED plan is the National Economic Development Plan.
[c]AEP is annual exceedance probability; 1/AEP is expected level of protection.
[d]Return period is calculated as 1 divided by the annual exceedance probability.

TABLE 7.1 Continued

90% Conditional Nonexceedance			FEMA Standard 3ft			95% Conditional Nonexceedance			90%-3ft-95% Combination		
Elev. (ft)	Freebd (ft)	Return period[d] (ft)	Elev. (ft)	Freebd (ft)	Return period[d] (ft)	Elev. (ft)	Freebd (ft)	Return period[d] (ft)	Elev. (ft)	Freebd (ft)	Return period[d] (ft)
(9)	(10)	(11)	(12)	(13)	(14)	(15)	(16)	(17)	(18)	(19)	(20)
43.4	1.8	240	44.6	3.0	400	**44.0**	**2.4**	**310**	44.0	2.4	310
48.5	2.4	190	**49.1**	**3.0**	**200**	52.3	6.2	240	49.1	3.0	200
31.5	2.3	400	32.2	3.0	780	**32.1**	**2.9**	**710**	32.1	2.9	710
796.6	1.3	5,000	798.3	3.0	10,000	**797.3**	**2.0**	**10,000**	797.3	2.0	10,000
834.3	2.9	170	**834.4**	**3.0**	**190**	835.2	3.8	380	834.4	3.0	190
910.7	1.5	200	912.2	3.0	1,000	**910.8**	**1.6**	**250**	910.8	1.6	250
1,330.9	4.6	**590**	1,329	.0	220	1,331.5	5.2	1,000	1,330.9	4.6	590
560.8	3.0	230	**560.8**	**3.0**	**230**	561.7	3.9	370	560.8	3.0	230
865.0	**3.3**	**210**	864.7	3.0	190	866.3	4.6	380	865.0	3.3	210
476.9	**5.8**	**910**	474.1	3.0	430	477.7	6.6	1,430	476.9	5.8	910
460.7	**5.6**	**200**	458.1	3.0	00	461.2	6.1	210	460.7	5.6	200
58.4	**3.5**	**210**	57.9	3.0	200	59.5	4.6	280	58.4	3.5	210
713.5	1.5	270	715.0	3.0	750	**713.9**	**1.9**	**290**	713.9	1.9	290
	3.0	230		3.0	230		4.0	370		3.3	250
	2.9	170		3.0	100		3.9	210		3.0	190
	1.3			3.0			1.6			1.6	
	5.8	5,000		3.0	10,000		6.6	10,000		5.8	10,000

Corps executes these instructions by estimating the heights of levees that will maximize economic benefits and are consistent with protecting the nation's environment (the National Economic Development (NED) alternative). By law, the Corps will build a levee to this height, with the federal government paying its share of the construction costs (according to the cost-sharing guidelines of the Water Resources Development Acts of 1986 and 1996). But if the community would like a levee higher than the NED levee, they must pay for *all* the additional construction costs. As Table 7.1 shows, the FEMA 3-feet-of-freeboard certification criterion required levee elevations greater than the NED elevation in 4 of the 11 communities in this group. If a community's levees were not certified, that community (if located in a zone subject to inundation by the 100-year flood) would remain in a Special Flood Hazard Area and would therefore be subjected to mandatory flood insurance purchase regulations, as well as relevant local land use regulations.

The 3-feet-of-freeboard concept was used as a design parameter to account for uncertainties associated with hydrologic and hydraulic analysis (Huffman and Eiker, 1991). If these uncertainties were accounted for, exceptions to the 3-feet-of-freeboard requirement were granted. One way to account for these uncertainties was to follow the Corps's risk analysis method. As documented in Appendix B, this gave rise in 1996 to a subsequent Corps–FEMA proposal for levee certification, which combined the 3-feet-of-freeboard requirement when risk analysis was *not* performed with a probabilistic estimate of protection using annual exceedance probability (AEP) when risk analysis *was* performed. The annual exceedance probability is calculated by determining the probabilities of flooding for flows of all possible exceedance probabilities (e.g., the 10-year flood, 50-year flood, 100-year flood, 200-year flood), then integrating over them to obtain a total probability of flooding from all floods. An expected value (i.e., the mean value) of the annual exceedance probability of one percent was selected as a decision criterion: if a levee had an expected annual exceedance probability of less than 1 percent, it was certified; otherwise, it was not.

In principle, the annual exceedance probability approach to levee certification enjoyed many benefits. It allowed a broader spectrum of uncertainties to enter the analysis than did the 3-feet-of-freeboard criterion. The data in Table 7.1 show that the NED elevation is greater than the 1 percent expected annual exceedance probability elevation for all communities in this group. If the Corps thus built to the NED elevation, all areas would be certified as satisfying the 1 percent annual exceedance probability criterion. No community would have to spend tens or hundreds of millions of dollars raising the elevation of their levees in order to be certified. Finally, the 1 percent expected annual exceedance probability criterion was simple to understand and to communicate.

But floodplain managers feared that the new annual exceedance probability criterion would not provide adequate protection against flooding. FEMA concluded, "for the 12 USACE projects [studied] the simulation exceedance (true) probability standard of 0.01 referenced in our April 1993 letter produced levee designs with only 0.1 to 1.5 feet of freeboard and contained the FEMA base flood with a reliability of between only 50 and 75%" (Krimm, 1996; see Appendix B). Thus, after three years, FEMA and the Corps rescinded the 1 percent expected annual exceedance probability criterion. In effect, they interpreted this criterion as something that should be satisfied with a high degree of assurance on each project, rather than just on the average over many projects.

Unfortunately, the risk analysis procedure at that time was embryonic. The annual exceedance probability itself has a range of uncertainty arising from uncertainties in the methods and data used to calculate it. By choosing the expected value of the annual exceedance probability, this range of uncertainty was not directly incorporated into the decision process. If the Corps had accounted for the various sources of uncertainty, as well as quantifying reasons for levee failure other than overtopping, the levels of freeboard and probability of containing a base flood would have been greater. In the committee's judgement, the major reasons for the difficulties with this 1 percent expected annual exceedance probability criterion were the Corps's method, the procedures that were followed to implement it, and the Corps's lack of experience (at that time) in using risk analysis.

CURRENT CERTIFICATION CRITERION

In March 1997 the Corps issued a new guidance circular (USACE, 1997e) based on the estimated probability that a levee would be able to pass a 100-year flood (the conditional nonexceedance probability; the circular is reproduced in Appendix B). The Corps considered three elevations for the levee: the elevation with a 90 percent conditional nonexceedance probability, a 95 percent conditional nonexceedance probability, and the elevation based on the old criterion of 3 feet of freeboard above the 100-year flood. The Corps and FEMA agreed to certify a levee if its elevation was at least (1) at the 90 percent conditional nonexceedance probability elevation, if the old criterion resulted in a conditional nonexceedance probability of less than 90 percent, (2) at the 95 percent conditional nonexceedance elevation, if the old criterion resulted in a conditional non-exceedance probability greater than 95 percent, or (3) at the old criterion's elevation, if the old criterion resulted in a conditional nonexceedance probability between 90 percent and 95 percent.

Figure 7.1 illustrates the levee certification decision tree. The FEMA level (FL) denotes the 100-year flood plus 3 feet of freeboard, and Corps Level 90 percent (CL90) and Corps Level 95 percent (CL95) denote the levels that have assurance (conditional nonexceedance probability) of 90 percent and 95 percent, respectively, of passing the 100-year flood. This mixture of the former 3-feet-of-freeboard approach and the reliability (90% or 95%) of passing the 100-year flood reflects a standard in most engineering designs—that is, a desired performance (lower limit) and avoidance of overdesign (upper limit).

This new certification criterion would require greater levee freeboard for East Peoria and other sites that formerly had lower likelihoods of passing a 100-year flood, and it would require less levee freeboard for West Sacramento and other sites that formerly had higher likelihoods of passing a 100-year flood. Like the criterion of 3 feet of freeboard, for some communities, the levee elevation required to satisfy the criterion is above the National Economic Development height. These communities would thus have to pay the entire cost of raising the levees to the certification height. In addition, the present criterion is awkward and confusing to the public.

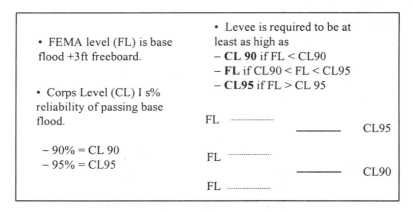

FIGURE 7.1 Current Corps–FEMA levee certification procedure.

ANALYSIS OF LEVEE SIZING CRITERIA

Data for 13 flood damage reduction projects in nine states were assembled in 1996 by the Corps's Hydrologic Engineering Center (HEC) in Davis, California (Table 7.1). No special effort was made to select particular kinds of projects or regions of the country in preparing this data set. The "base elevation" in Column (1) is the water surface elevation above mean sea level of the median 100-year flood discharge. For each levee sizing alternative, three data values are shown: (1) the levee elevation, (2) the corresponding freeboard above the base elevation, and (3) the inverse of the expected annual exceedance probability for this elevation, which is a measure in years of the expected return period between failures of the levee. This value is sometimes called the levee's "level of protection." To acknowledge that this quantity is a random variable with a range of variation, it is called here the "expected level of protection."

Data for six methods of levee sizing are presented in Table 7.1: (1) the National Economic Development (NED) plan level, (2) the 1 percent annual exceedance probability level (the 1993–1996 certification criterion), (3) the 90 percent conditional nonexceedance probability level (CL90 in Figure 7.1), (4) the FEMA standard level (base elevation plus 3 feet; FL in Figure 7.1), (5) the 95 percent conditional nonexceedance probability level (CL 95 in Figure 7.1), and (6) the current Corps–FEMA certification level combining the CL 90, FL, and CL95 values (90%-3ft-95%). Summary statistics of the freeboard and level of protection provided by each alternative are provided at the bottom of Table 7.1.

The freeboard required for levees having an expected annual exceedance probability of 1 percent averages 0.9 feet and ranges from 0.1 to 3.2 feet (Column 7). The discrepancy between these values and the traditional three feet of freeboard led the Corps and FEMA in 1996 to abandon the mean 1 percent expected annual exceedance probability criterion.

Columns (9), (12), and (15) of Table 7.1 show the three calculated levee elevations associated with the new criterion for the 13 communities. Column (9) indicates the elevation with a 90 percent likelihood of passing the 100-year flood. Column (12) indicates the 100-year flood elevation plus 3 feet of freeboard. Column (15) indicates the elevation with a 95 percent likelihood of passing the 100-year flood. In each case, bold type indicates the levee height required for certification, shown in Column (18). The combined criterion agreed to by the Corps and FEMA results in a certification level determined by 3 feet of freeboard for three communities, by a 90 percent chance of passing the 100-year flood for five communities, and by a 95 percent chance of passing the 100-year flood for five communities. Thus, for five communities, the likelihood of the levees passing the 100-year flood is 90 percent, for three communities, it is between 90 percent and 95 percent, and for five communities, it is 95 percent. However, the seemingly close agreement among the three criteria concerning levee height masks large cost differences in modifying levees to attain these heights.

The statistics summarized at the bottom of Table 7.1 show that:

• the FEMA standard of 3 feet of freeboard provides a median expected level of protection of approximately 230 years, with a range of <100 years to >10,000 years,
• the Corps–FEMA 90%-3ft-95% criterion provides an average of 3.3 feet of freeboard and yields a median expected level of protection of approximately 250 years, with a range of 190 to 10,000 years,
• the 90 percent conditional nonexceedance probability provides an average of 3.0 feet of freeboard, and a median expected level of protection of approximately 230 years, with a range of 170 to 5,000 years, and,
• the 95 percent conditional nonexceedance probability provides an average of 4.0 feet of freeboard and a median expected level of protection of approximately 370 years, with a range of 210 to 10,000 years.

These data indicate that over the range of these 13 projects, the current Corps–FEMA criterion is slightly more conservative than the FEMA standard 3-feet criterion, and that the 95 percent conditional nonexceedance probability level is significantly more conservative. Levees designed to the 90 percent conditional nonexceedance probability level have approximately the same median level of safety as the traditional standard of 3 feet of freeboard.[2]

The range in the freeboard required for the Corps–FEMA combined criterion (90%-3ft-95%) compared to the FEMA standard 3-feet criterion

---

[2] The risk analysis method adopted by the Corps involves a process that is considerably more complicated and requires much more data and analysis than the former procedure of adding 3 feet of freeboard. Is it possible that some type of a simple freeboard requirement would be a good approximation to the risk analysis results? Columns (10) and (16) of Table 7.1 show the amount of freeboard that would be required for each site to pass the 100-year flood with likelihoods of 90% and 95%. For a likelihood of 90%, the freeboard requirement ranges from 1.3 feet to 5.8 feet. For a likelihood of 95%, the freeboard requirement ranges from 1.6 feet to 6.6 feet. No simple freeboard measure gives a good approximation to the risk analysis measure.

is illustrated in Figure 7.2. It can be seen that the current Corps-FEMA combined criterion requires freeboard ranging from 1.6 feet in Hamburg, Iowa, to 5.8 feet in Sny Island Levee Drainage District (LDD), Illinois. For most projects on the right of the figure, the 90 percent conditional nonexceedance probability level defines the certified elevation. On the left of the figure, there are several projects where either the 3-feet or 95 percent criterion prevails, adding approximately 0.6 feet of freeboard to the 90 percent level. It is clear from Figure 7.2 that except for American River, California, the range in required freeboard between the 90 percent and 95 percent values is small when compared to the variation in these elevations from project to project. To reiterate, however, "small" differences in levee height can translate into large differences in the cost of the levees.

Figure 7.2 is an impressive demonstration of the value of risk analysis in assessing the required height of levees. It shows that to meet a consistent national standard of levee safety, as little as 2 feet of freeboard may be required in one location, while as much as 6 feet may be required in another. This raises serious questions regarding the degree of safety being provided by the current FEMA standard freeboard of 3 feet required for non-Corps projects. It shows that the traditional 3 feet of freeboard masks a significant degree of variation of risk of failure in levees built to this standard for the citizens protected by these levees. This variation in risk of failure can be quantified by the Corps's risk analysis procedure.

Figure 7.3 compares the elevation required for the National Economic Development plan with that for the three levee certification criteria for 11 of the 13 levee projects in Table 7.1 for which an National Economic Development plan elevation exists. On average, the National Economic Development plan provides approximately 0.5 feet of additional freeboard beyond the Corps-FEMA 90%-3ft-95% criterion, but in 4 of the 11 projects (Portage, Wisconsin; Pender, Nebraska; Guadalupe River, Texas; White River, Indiana), the National Economic Development plan would not provide sufficient elevation for the levee to be certifiable. In these cases, the Corps's local partners would be required to pay the entire cost of raising the levee to the certifiable level. Few communities would likely be able or willing to do this. As a result, the levees will not be certified, part or all of the community will remain in the Special Flood Hazard Area (SFHA) and will thus be subject to mandatory flood insurance purchase guidelines—and to generally higher flood insurance premiums than communities located outside of SFHAs (loca-

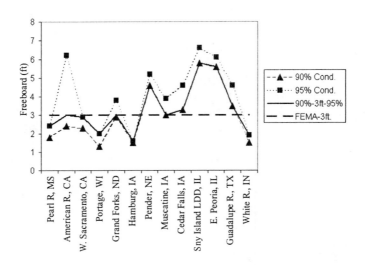

FIGURE 7.2 Current Levee Certification Criterion (90%-3ft-95%) and the FEMA Standard Three Feet.

tion within an SFHA may also subject a community to more stringent land use regulations). With higher premiums, fewer people are likely to purchase flood insurance and, when the inevitable flood occurs, they may face financial ruin. The federal government will, consequently, have little choice but to offer grants and low-interest loans, which is precisely what Congress wanted to eliminate by establishing the National Flood Insurance Program.

There are two consequences of using an annual exceedance probability of 1/230 rather than 1/100 as the certification criterion. The first has to do with whether the flood insurance premiums are actuarially fair. If the premiums are based on floods with a probability of 1/100 and the actual probability is 1/230, the premiums are far too high. These premiums would thus discourage people from buying flood insurance, which is the opposite of what Congress desired. The second is that Congress gave benefits to communities whose levees are certified, benefits such as lower insurance rates and an exemption from having to buy flood

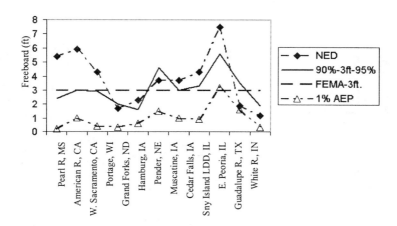

FIGURE 7.3 Comparison of freeboard requirements for the National Economic Development plan elevation and those arising from the levee certification criteria.

insurance. If the certification criterion is more stringent than Congress desired, some communities will have to pay higher insurance rates and will have to buy flood insurance that they need not have purchased.

The data for the 13 flood damage reduction projects in Table 7.1 are plotted in Figure 7.4 to illustrate the relationship between levee freeboard and the expected level of protection for a number of levee sizing criteria. The expected level of protection is the inverse of the expected value of the annual exceedance probability. As anticipated, the expected level of protection increases with the amount of freeboard required. However, the data for the Portage, Wisconsin, flood damage reduction project in the dashed box on the lower right of the diagram represent an anomaly. These values indicate that the expected level of protection provided by about 2 feet of freeboard at Portage is much higher at this location than in the other 12 projects surveyed. This anomaly may be a result of some peculiarity in the topography at Portage, or the risk analysis for this project may have been done differently for this project than for the other projects.

Part of the stimulus for this committee's establishment arose from a conflict among the Corps and state and local interests in Wisconsin con-

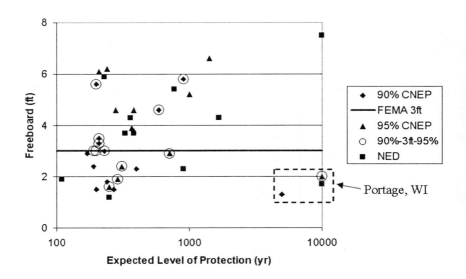

FIGURE 7.4 The expected level of protection provided by levee free-board.

cerning the appropriate levee height for the Portage project. Figure 7.4 suggests that the Portage project is not representative of the general relationship between freeboard and expected level of protection provided by Corps projects. According to the Corps's risk analysis, requiring 3 feet of freeboard on the levee at Portage led to an abnormally expensive project. Assuming that the Corps's risk analysis was correct, the Corps was correct in resisting so expensive a project that went beyond the levee called for in the National Economic Development alternative.

All Corps flood damage reduction studies prepared by Corps offices across the country are reviewed at Corps Headquarters in Washington, D.C. As part of that review process, the committee recommends that the Corps maintain an inventory from past projects of the amount of freeboard provided for the base flood and of the resulting expected level of protection provided by project levees.

To summarize, the former 3-feet-of-freeboard criterion for certifying levees (1) provided quite different levels of flood protection to different communities, (2) was unnecessarily expensive in some communities, and (3) was above the National Economic Development elevation in some communities, requiring these communities to pay for all the costs of the additional levee elevation. The new certification criterion is an improvement, as it results in a narrower range of protection (90% to 95% chance of passing a 100-year flood, rather than 45% to 99.9%). The new criterion, however, does not solve the problems associated with the former 3-feet-of-freeboard criterion.

One possible certification criterion based on risk analysis is the use of a conditional nonexceedance probability for the 100-year flood that must be at least some appropriate value, such as 0.90. This criterion focuses on the best estimate of the 100-year flood and can be used in both unprotected areas (delineation of the floodplain with some extra safety factor) and areas protected by levees. However, this measure is unnecessarily complicated and difficult to understand.

In summary, and based upon this review, the committee recommends that the Corps and FEMA promptly develop an improved levee certification procedure. The committee recommends a roughly equivalent criterion be used: the annual exceedance probability (the probability that the area will be flooded during a year). The committee recognizes that considerable resources and time will be required to modify and implement this recommendation. In this interim period, the committee therefore recommends that the Corps and FEMA adopt a single conditional nonexceedance probability for use in their joint levee certification program.

A sample of 13 projects is of limited size for drawing general conclusions about the equivalence of risk measures among different levee sizing criteria. The committee therefore also recommends that the Corps and FEMA compile comparable data to those in Table 7.1 for a significantly larger number of levee projects, with a view to replacing the current 90%-3ft-95% levee certification rule with a rule based on the annual probability of flooding.

TECHNICAL CORRECTIONS TO THE
CURRENT CERTIFICATION PROCEDURE

The current Corps–FEMA levee certification procedure is an important step toward implementing a rational approach to assessing the degree of flood protection provided by a levee system. The procedure accounts for uncertainty in assessing the frequency and severity of future flooding, and it estimates the probability that the levee system will perform as intended. The committee recognizes the difficulties the Corps and FEMA face in attempting to establish a rational, workable procedure for certifying levee systems. The Corps and FEMA are to be commended for deciding to directly face these challenges.

In reviewing the current certification procedure, the committee identified four technical issues that should be addressed in the near term to improve the risk analysis method. First, the procedure does not use straightforward probabilistic measures, such as annual probabilities of flooding, that are easily interpreted and easily compared to program mandates. The procedure is difficult for the public, and for the informed technical community, to understand and to communicate.

Second, the current certification procedure examines the base (100-year) flood alone and not the range of floods that can occur. For example, the process does not consider the potential that the levee system could fail due to wind and waves leading to overtopping, embankment failure, or other related factors during the 90-year event. Wind, waves, debris, and ice could lead to the 90-year flood overtopping the levee, or absence of them could lead to the 110-year flood not overtopping the levee. The current procedure thus gives a probability of flooding, assuming that the 100-year flood occurs. A more relevant measure would be the probability of flooding, given the distribution of floods of all sizes.

Third, evaluation of the uncertainties in levee performance does not comprehensively address sources of knowledge uncertainty in the geotechnical evaluation of levee system reliability, river hydraulics, and foundation or embankment failure; thus, an arbitrary level of reliability is being evaluated as part of the certification process.

Fourth, the current procedure focuses on the portion of each levee that is most likely to fail, which may not provide a sufficient analysis of the performance of levees as a system. Assessment of the levee system should account for the potential for failure at any point along the levee during a flood event, considering multiple modes of levee failure—including overtopping, piping, embankment instability, foundation instability, and other geotechnical considerations. These technical corrections do not require a review of policy implications or cost. They could therefore be implemented immediately.

AN ALTERNATIVE CRITERION:
A LONGER-TERM CHANGE

Beyond the improvements to the current procedure suggested above, the Corps and FEMA should work toward an alternative criterion based on a simple "annual probability of levee system failure (annual exceedance probability)" accounting for both natural variation and knowledge uncertainty and accounting for threats to the levee system other than just overtopping. Levee system failure is equated here with the failure to prevent inundation of any area that should be protected (i.e., the expected level of protection). Such a criterion would have direct benefits compared to the current criterion: (1) there would be a direct measure of the level of protection, (2) the criterion would be focused on protection from all floods, not just on the 100-year flood, and (3) the criterion would be easier for the public, Congress, and the engineering community to understand.

The committee believes that earlier difficulties encountered with an annual-probability-of-flooding approach can now be overcome because of the Corps's greater experience with risk analysis and because the Corps can now consider target annual probabilities in light of historical practice. The criterion for certifying a levee should be that it provides adequate protection against failure of the flood damage reduction system. Because the old criterion for levee certification produced different levels of flood protection for different communities, there is no reason to replicate its levels of protection in the new criterion.

The former certification criterion of 3 feet of freeboard provided an expected annual exceedance probability of roughly 0.00435 (return period of 230 years), while the 90 percent nonexceedance probability levee provided the same expected annual exceedance probability of 0.00435. The committee recommends that levees provide a uniform level of protection across communities. It is not obvious which level of protection should be chosen. In the committee's judgment, the criterion should aim to provide the level of protection provided to the most people in the past—the median level historically provided. Based on a small sample of Corps flood damage reduction projects, that median annual exceedance probability is roughly 1/230.

The committee also recommends that the Corps develop a table showing percentiles of variability in the annual exceedance probability or showing levels of protection similar to that used for the measures of economic performance in project planning. By choosing an appropriate percentile value in this range, a corresponding level of assurance can be obtained that the expected level of protection is at least 100 years, as required. It was the lack of allowance for this variability that led to the abandonment of the annual exceedance probability criterion during the 1990s.

A difficulty with establishing a criterion based on an expected annual exceedance probability of 1/230 (0.00435) is that the resulting elevation often exceeds the elevation of the National Economic Development (NED) alternative levee. Few if any of these communities would elect to spend tens or hundreds of millions of dollars to raise their levees above the NED elevation to become certified. However, the judgment regarding the level at which to certify levees is a political, not a technical judgment.

The alternative criterion changes the focus from the 100-year flood to the expected probability of flooding. Although the committee judges the latter to be what Congress intended in the National Flood Insurance Act of 1968 and its later amendments, we recognize that it has not been the criterion used in federal regulations. It is therefore recognized that review and discussion will likely be required before implementing the alternative criterion. Although shifting the focus to the annual probability of flooding is the more desirable alternative, the committee does not want to delay implementation of the technical corrections. All of these corrections are required for implementing the alternative criterion. Thus, nothing is wasted by immediate implementation of the corrections. At the same time, the committee recommends the alternative criterion, recognizing that its implementation will require examination and discussion.

8

Conclusions and Recommendations

The Committee on Risk-Based Analyses for Flood Damage Reduction embarked upon this study assuming that it would produce a technical report regarding the application of risk analysis within the Corps of Engineers's flood damage reduction studies. The charge to the committee was:

> The Secretary (Army) shall enter into an agreement with the National Academy of Sciences to conduct a study of the Corps of Engineers' use of risk analysis for the evaluation of hydrology, hydraulics, and economics in flood damage reduction studies. The study shall include—
>
> a) an evaluation of the impact of risk-based analysis on project formulation, project economic justification, and minimum engineering and safety standards; and
> b) a review of studies conducted using risk-based analysis to determine—
>> i) the scientific validity of applying risk-based analysis in these studies; and
>> ii) the impact of using risk-based analysis as it relates to current policy and procedures of the Corps of Engineers.

As the committee proceeded with its study and its discussions, it learned that the issues of risk analysis within the Corps were quite complicated, and that a full understanding of these methods and their applications required a broader investigation than the committee anticipated at the outset. To adequately address its charge, the committee ultimately examined several topics in addition to its technical analysis of risk-based techniques.

Thus, while this study presented several challenges, it also offered a unique opportunity to analyze a host of critical, emerging issues in statistical hydrology, geotechnical engineering, economics, communications, and public policy. In this study, these multiple disciplines, and the committee members representing them, were bound by common interests in risk analysis, floods, Corps of Engineers practices, and floodplain management.

The Corps of Engineers's adoption of risk analysis procedures that explicitly recognize and quantify hydrologic, hydraulic, geotechnical, and economic uncertainties should lead to projects better tailored to local conditions and to available information, thus better achieving social objectives and flood damage reduction goals. For example, the new procedures represent a significant advance over traditional levee freeboard requirements that led to inconsistent levels of flood protection among projects. The replacement of this long standing approach with new risk analysis methods required significant conceptual and methodological development. **The new techniques are a significant step forward and the Corps should be greatly commended for embracing contemporary, but complicated, techniques and for departing from a traditional approach that has been overtaken by modern scientific advances**. There should be no turning back from this important step forward.

It bears repeating that the former levee freeboard standard did not provide consistent levels of flood protection across the nation. A consistent protection standard must properly account for local and regional differences in topography, hydrology, and hydraulics, which the standard freeboard approach did not. For instance, as little as 2 feet of freeboard may be required to provide adequate flood protection in some areas, while in others, as much as 6 feet may be required. The traditional freeboard standard masks a significant degree of variation of risk of levee failure for citizens protected by these levees. This variation in risk of failure can be quantified by the Corps's new risk analysis procedure.

While the Corps is to be strongly commended, it is important that the Corps promptly address and resolve issues identified in this report in order for risk analysis methods to eventually become well founded, well documented, and correctly executed. Risk analysis methods need to be clearly understood and clearly documented, and results effectively communicated to the public, its elected officials, other federal agencies, the U.S. Congress, and to the engineering community.

RISK ANALYSIS TECHNIQUES

The committee reviewed the Corps's risk-based applications, including the Hydrologic Engineering Center's Flood Damage Assessment (HEC-FDA) computer program for calculating flood damage risk. The new method builds upon the deterministic approach to flood damage estimation that evolved over decades of use. The new method benefits from these decades of experience, but also suffers from the difficulty of translating deterministic practice into a probabilistic framework. While the Corps is to be praised for its adoption and implementation of the risk analysis methods, several technical issues remain only partially addressed. Particular concerns relate to models of hydrologic, hydraulic, and geotechnical uncertainties, performance metrics, and the economic uncertainty analysis.

Risk Measures and Modeling

According to the U.S. federal *Principles and Guidelines*, estimates of expected annual damages (EAD) are the primary criterion (performance metric) for project selection. These are supplemented by the annual exceedance probability (AEP), which describes the likelihood of flooding in areas that should be protected. These two project performance metrics are important and appropriately summarize economic and safety dimensions of system performance. They adequately characterize the performance of flood damage reduction projects.

For levee certification and engineering purposes, it is useful to calculate other system reliability measures, such as the conditional nonexceedance probability for the 1 per cent flood (this describes the uncertainty in the water height of the flood with a 0.01 probability of occurring in a given year). In the committee's view, however, such engineering reliability measures, which are difficult to understand, should not be used to communicate flood risk to the public. Conditional nonexceedance probabilities are an awkward combination of traditional and new risk-based concepts and are easily misunderstood. The concept of annual exceedance probability is clearer and provides the needed information. **The committee thus recommends that the Corps use *annual exceed-***

ance probability **as the performance measure of engineering risk.**

The construction of risk measures rests upon complete and accurate specification of the uncertainties in each component of an analysis, and upon correct implementation of procedures to: (1) estimate knowledge uncertainties, (2) incorporate those uncertainties in the risk analysis, and (3) properly propagate the uncertainties in individual parts of the analysis to the final results. **As the current method has shortcomings in these areas, the committee recommends that the Corps improve its analysis of economic, hydrologic, hydraulic, and geotechnical engineering uncertainties.**

The committee reviewed the Corps's risk analysis model and the HEC-FDA computer code used to perform calculations. A concern is whether all of the important uncertainties are included in the analysis and if those uncertainties are appropriately represented. The Corps's method should clearly distinguish between natural variability and knowledge uncertainty. In some cases, such as stage-discharge relations, uncertainties due to natural variability appear to be incorrectly subsumed within knowledge uncertainties; in other cases, such as geotechnical levee performance, knowledge uncertainties are incorrectly subsumed within natural variability.

The committee thus recommends that the Corps focus greater attention on the probabilistic issues of identifying, estimating, and combining uncertainties. This recommendation is important because, in the way the analyses are conducted, knowledge uncertainties (lack of scientific understanding of events and processes) explicitly affect project performance measures, but natural variability does not.

The committee recommends that the Corps strive to reduce the considerable variation in the estimates of water surface elevation when using different models of river hydraulics. The Corps's experiences in applying alternative methods to estimate flood stage indicate that there can be substantial differences in the results.

The committee recommends that the Corps's risk analysis method evaluate the performance of a levee as a spatially distributed system. The geotechnical evaluation of a levee, which may be many miles long, should account for the potential for failure at any point along the levee during a flood. Multiple modes of failure (e.g., overtopping, embankment instability), correlation of embankment and foundation properties, and the potential for multiple levee section failures during a flood should also be considered. The current procedure treats a levee within each damage reach as independent and distinct from one reach to the next. Further, within a reach, the analysis focuses on the portion of each levee that is most likely to fail. This does not provide a sufficient analysis of the performance of the entire levee. This has important implications for not only geotechnical and economic analysis of flood damages, but also for levee certification.

The Corps's new geotechnical reliability model would benefit greatly from field validation. The nation has many years of experience with levee performance and, unfortunately, also with levee failures. Much of this experience is documented and accessible to federal agencies. **The committee recommends that the Corps undertake statistical ex post studies to compare predictions of geotechnical failure probabilities made by the reliability model against frequencies of actual levee failures during floods. The committee also recommends that the Corps conduct statistical ex post studies with respect to the performance of other flood damage reduction structures. These latter studies should be conducted in order to identify vulnerabilities (failure modes) of these systems and to verify engineering reliability models.**

Economics

A key innovation of the Corps's new risk analysis procedures is the derivation of the uncertainty in economic benefits due to knowledge uncertainties. This analysis is performed using a Monte Carlo procedure that evaluates expected annual damages using different possible combinations for hydrology, hydraulics, geotechnical and economic model parameters. This uncertainty analysis is currently performed as if the errors for each individual structure that might suffer damage are independent of one another. This fails to include the correlations and interaction among these parameters and thus misrepresents the uncertainty in damages in individual reaches, and for the project as a whole. For example, if a house has a higher economic value than estimated, or is surveyed to be at too low an elevation, similar houses on the same block are likely to experience similar errors. Unfortunately, the current Corps procedures neglect these sources of correlation and as a result misrepresent the overall uncertainty in project damages and benefits.

For similar reasons, the Corps's procedure incorrectly computes the uncertainty associated with differences in benefits from alternative projects components. The current practice of summing and subtracting percentiles of uncertainty distributions for damages in different reaches, and for alternative project components, is statistically incorrect unless there is perfect correlation among the values. This is not the case, although there are strong interrelationships. As a result, the uncertainty assigned to the computed annual project damages and benefits is incorrect.

The assessment of uncertainty in economic performance could be significantly simplified to yield more precise measures of uncertainty by focusing on uncertainty in project benefits for each individual structure, while considering the interrelationships among those uncertainties instead of accumulating damages over damage reaches. What is critical is not accurate assessment of flood damage, but rather accurate assessment of the degree to which the project will reduce flood damage. If the flood damage to each structure in the floodplain is estimated both with and without the project plan, the benefit provided by the plan to this structure is the amount of reduced damage. Errors in estimating flood damage arise from errors in flood discharge, water surface elevation and the structure's economic value. These errors affect the damages both with and without the plan in nearly the same way; thus, when the benefit is calculated, the effects of these errors may nearly be cancelled out. The uncertainty in estimation of project benefits accumulated over all structures could thus be significantly less than the uncertainty in estimating project damage. Moreover, the impact of correlation among the analysis errors may be similarly diminished. Detailed study of alternatives to the current method of accounting for uncertainty in economic performance is needed to identify more efficient and statistically robust procedures.

The committee recommends that the Corps calculate the risks associated with flooding, and the benefits of a flood damage reduction project, structure by structure, rather than conducting risk analysis on damage aggregated over groups of structures in damage reaches.

CONSISTENT TERMINOLOGY

The committee recommends that the Corps adopt and use a consistent vocabulary for distinguishing among natural variability, knowledge uncertainty, and measures of system reliability. Although difficult to assess, the distinction between natural variability and knowledge uncertainty is important because of the different roles these sources of uncertainty play in risk analysis.

LEVEE CERTIFICATION

In the early 1990s the Corps adopted a risk analysis approach to re-place the practice of certifying levees that had 3 feet of freeboard above the 100-year flood. This former criterion provided different levels of flood protection to different sites. Both the new risk analysis approach and the levels of flood protection it provided were controversial. Nego-tiations with the Federal Emergency Management Agency (FEMA) led to the current practice of certifying a levee based on a three-tiered deci-sion rule, using either 3 feet of freeboard, or a 90 percent conditional nonexceedance probability of passing a 100-year flood, or a 95 percent conditional nonexceedance probability. This criterion is a step forward and a reasonable transition from the older levee certification criterion into the new risk analysis framework. It has problems, however: (1) it still leads to different levels of flood protection for different projects, (2) the three-tiered decision rule is unnecessarily complicated, (3) the method evaluates levees individually and not as a system, and (4) certifi-cation considers only the 100-year flood, not the full range of floods.

The committee recommends that the federal levee certification program focus not upon some level of assurance of passing the 100-year flood, but rather upon "annual exceedance probability." This is the probability that an area protected by a levee system will be flooded by any potential flood. **This annual exceedance probability of flood-ing should include uncertainties derived from both natural variabil-ity and knowledge uncertainty.**

The adoption and implementation of annual exceedance probability as the key criterion in levee certification will require significant re-sources and time. **Until the measure of annual exceedance probability approach is adopted as the key criterion for levee certification, the committee recommends that the Corps and FEMA adopt a single conditional nonexceedance probability for levee certification.**

The former certification criterion was flawed in that it produced vastly different levels of flood protection for different communities. **The committee recommends that the certification criterion provide a uni-form level of flood protection.** Which level of protection to choose is not obvious. Insisting on the highest level of protection would mean that only a very small proportion of levees would be certified. **In the com-mittee's judgment, the certification criterion should be the level of protection provided to most people in the past—the median level his-torically provided.** Based upon a small sample of all Corps flood dam-age reduction projects, the committee found that the median annual exceedance probability of Corps flood damage reduction projects is approximately 1/230.

This is the committee's best estimate of the median annual exceedance probability. **To obtain a more reliable measure of the median annual exceedance probability of approved projects, the committee recommends that the Corps examine a larger number of flood damage reduction projects and audit the process of estimating the annual exceedance probability for these projects.**

The committee also recommends that the Corps develop a table showing percentiles of variability in the annual exceedance probability of its flood damage reduction projects. By choosing an appropriate percentile value in this range, a corresponding level of assurance can be obtained that the expected level of protection is at least 100 years, as required. The lack of allowance for this variability led to the abandonment of the annual exceedance probability criterion during the 1990s.

FLOODPLAIN MANAGEMENT

Neither the U.S. Congress nor the Corps of Engineers have identified an explicit goal for management of the nation's floodplains. The committee is of the view that the goal of floodplain management should be to use the land for the greatest social benefit. Broadening the scope of Corps risk analysis and expanding the types of alternatives considered would provide more useful insight about how to best achieve this goal.

If conducted using the best information available, risk analysis can provide substantial insight for making floodplain management decisions. The Corps's risk analysis method quantitatively considers only direct damage reductions and costs of alternatives. It does not consider potential loss of life or environmental and social consequences. The benefits that can be claimed within Corps water resources project planning studies are specified by the federal *Principles and Guidelines* and not by the Corps.

To ensure that the Corps's flood damage reduction projects provide adequate social and environmental benefits, the committee recommends that potential loss of life, other social consequences, and environmental consequences be explicitly addressed in the Corps's risk analysis. This will improve the ability to make informed decisions about floodplain management and should ultimately lead to better use of our nation's floodplains. Floodplain management alternatives should not be limited to structural alternatives such as levees, dikes, and dams. Information and communication alternatives, such as warning systems and zoning regulations, should be considered both separately and in conjunction with structural alternatives. The committee recognizes that the Corps alone cannot implement these recommendations, nor is such a broadening of the risk analyses likely to occur over a short period of time.

To appropriately include such consequences and their relative importance, the committee recommends that the ecological, health, and other social effects of Corps flood damage reduction studies, and the tradeoffs between them, be quantified to the extent possible and included in the National Economic Development Plan. More explicit efforts at including these types of consequences and values in the Corps's benefit–cost calculations should increase social benefits of the Corps's flood damage reduction studies. Examples of these consequences that are not included in the current benefit–cost guidelines contained within the *Principles and Guidelines* include lives saved (by structural and nonstructural projects), damages avoided to structures in floodplain evacuation projects, and preservation of biodiversity. Appropriate revisions of existing legislation, consistent with this recommendation, may have to be enacted by the U.S. Congress. The Corps should seek guidance from the Office of Management and Budget and seek consistency with other federal agencies on the use of alternative metrics for incorporating potential loss of life, environmental impacts, and other effects of floods.

The U.S. Army Corps of Engineers is a national and international leader in addressing flood-related problems. The committee applauds the agency's commitment and applauds Corps personnel in adopting, developing, and implementing the risk analysis approach. It is imperative that the Corps take the time needed to complete this new approach so that it achieves its potential. This new approach represents a significant step forward for the Corps, and for flood damage reduction studies, in general.

The committee commends the Corps for recognizing the limitations of the former procedures and for having the courage to undertake the development of a new, controversial, and technically difficult risk analysis approach. We offer this report in the spirit of constructive advice and in hopes of promoting wiser use of the nation's floodplains.

References

Al-Futaisi, A., and J. Stedinger. 1999. Integrating hydrologic and economic uncertainties into flood risk management project design. Journal of Water Resources Planning and Management 125(6):314-324.

Bazovsky, I., 1961. Reliability Theory and Practice. Englewood Cliffs, N.J.: Prentice-Hall.

Beard, L. R. 1962. Statistical Methods in Hydrology. Revised from an earlier edition published in 1952 under Civil Works Investigations Project CW-151 by the U.S. Army Engineer District. Sacramento, Calif.: U. S. Army Corps of Engineers.

Bobee, B. and F. Ashkar. 1991. The Gamma Family and Derived Distributions Applied in Hydrology. Littleton, Colo.: Water Resources Publications.

Brown, C. A., and W. J. Graham. 1988. Assessing the threat to life from dam failure. Water Resources Bulletin 24:1303-1309.

Budnitz, R. J., G. Apostolakis, D. M. Boore, L. S. Cluff, K. J. Copersmith, C. A. Cornell, and P. A. Morris. 1997. Recommendations for Probabilistic Seismic Hazard Analysis: Guidance on Uncertainty and Use of Experts. Prepared for the U.S. Nuclear Regulatory Commission, NUREG/CR-6372. Washington, D.C.: U.S. Nuclear Regulatory Commission.

Calabro, S. R. 1962. Reliability Principles and Practices. New York, N.Y.: McGraw-Hill.

Carhart, R. R. 1953. A Survey of the Current Status of the Electronic Reliability Problem. Research Memo RM-1131. Santa Monica, Calif: Rand Corporation.

Chow, V. T., D. R. Maidment, and L. W. Mays. 1988. Applied Hydrology. New York, N.Y.: McGraw-Hill.

Chowdhury, J. U., and J. R. Stedinger. 1991. Confidence intervals for design floods with estimated skew coefficient. Journal of Hydraulic Engineering 117(7):811-831.

Cornell, C. A. 1968. Engineering seismic risk analysis. Bulletin of the Seismological Society of America 58:1583-1606.

Cornell, C. A., and E. H. Vanmarcke. 1969. The major influences on seismic risk. Pp. 69-93 in Proceedings of the Fourth World Conference on Earthquake Engineering. Santiago, Chile.

Cressie, N. A. 1991. Statistics for Spatial Data. New York, N.Y.: John Wiley.

Davis, D. W. 1991. A risk and uncertainty based concept for sizing levee projects. Pp. 231-249 in Proceedings of a Hydrology and Hydraulics Workshop on Riverine Levee Freeboard. Davis, Calif.: U. S. Army Corps of Engineers Hydrologic Engineering Center.

Fischhoff, B., S. Lichtenstein, P. Slovic, S. L. Derby, and R. L. Keeney. 1981. Acceptable Risk. Cambridge, Mass.: Cambridge University Press.

Fishburn, P. C. 1970. Utility Theory for Decision-Making. New York, N.Y.: John Wiley.

Gumbel, E. J. 1941. The return period of flood flows. The Annals of Mathematical Statistics 12(2):163-190.

Haimes, Y. Y. 1998. Risk Modelling, Assessment and Management. New York, N.Y.: John Wiley.

Huffman, R. G., and E. Eiker. 1991. Freeboard design for urban levees and floodwalls. Pp. 5-11 in Proceedings of a Hydrology and Hydraulics Workshop on Riverine Levee Freeboard, Monticello, Minnesota. Davis, Calif.: U.S. Army Corps of Engineering Hydrologic Engineering Center.

Hershfield, D. M. 1961. Rainfall frequency atlas of the United States for Durations from 30 Minutes to 24 Hours and Return Periods from 1 to 100 Years. Technical Paper 40. Washington, D.C.: US Dept of Commerce, Weather Bureau.

Hydrologic Engineering Center. 1998a. HEC-FDA Flood Damage Reduction Analysis, Users Manual, Version 1.0. Davis, Calif.: US Army Corps of Engineers, Hydrologic Engineering Center.

Hydrologic Engineering Center. 1998b. HEC-HMS Hydrologic Modeling System, Users Manual, Version 1.0. Davis, Calif.: U.S. Army Corps of Engineers, Hydrologic Engineering Center.

IACWD (Interagency Advisory Committee on Water Data). 1981. Guidelines for Determining Flood Flow Frequency, Hydrology Subcommittee, Bulletin # 17B (Revised and Corrected). Reston, Vir.: U.S. Geological Survey.

IFMRC (Interagency Floodplain Management Review Committee). 1994. Sharing the Challenge: Floodplain Management into the 21st Century. Report of the Interagency Floodplain Administration Floodplain Management Task Force. Washington, D.C.: U.S. Government Printing Office.

Institute of Hydrology. 1999. Flood Estimation Handbook. 5 Vols. Wallingford, United Kingdom: Institute of Hydrology.

Karl, T. R., R. W. Knight, D. R. Easterling, and R. C. Quayle. 1996. Indices of climate change for the United States. Bulletin of the American Meteorological Society 77 (2):279-292.

Keeney, R. L. 1992. Value-Focused Thinking. Cambridge, Mass.: Harvard University Press.

Keeney, R. L., and H. Raiffa. 1976. Decisions with Multiple Objectives. New York, N.Y.: John Wiley. Republished in 1993. New York, N.Y.: Cambridge University Press.

Keeney, R. L., T. L. McDaniels, and V. L. Ridge-Cooney. 1996. Using values in planning wastewater facilities for Metropolitan Seattle. Water Resources Bulletin 32:293-303.

Krimm, R. 1996. Letter from Richard W. Krimm, Federal Emergency Management Agency to Major General Stanley G. Genega, U.S. Army Corps of Engineers. March 21, 1996.

Kusler, J., and L. Larson. 1993. Beyond the ark: A new approach to U.S. floodplain management. Environment 35(5):7-11, 31-34.

Lave, L., D. Resendiz-Carillo, and F. McMichael. 1990. Safety goals for high-hazard dams: Are dams too safe? Water Resources Research 26:1383-1391.

Lesher, M., and P. Foley. 1997. Risk-based analysis for evaluation of alternatives: Grand Forks, North Dakota and Crookston, Minnesota. In Proceedings of Hydrology and Hydraulics Workshop on Risk-based Analysis for Flood Damage Reduction Studies. Davis, Calif.: U.S. Army Corps of Engineers Hydrologic Engineering Center

Marshall and Swift. 1999. Residential Cost Handbook, Marshall and Swift Co. http://www.marshallswift.com/realestateproductpages/RCH.htm (limited access).

McCuen, R. H. 1979. Map skew. Journal of the American Society of Civil Engineers Water Resources Planning and Management Division 105 (WR2): 269-277 (with Closure 107 (WR2:582, 1981)).

Morgan M. G., B. Fischhoff, A. Bostrom., C. Atman, and L. Lave. 1992. Communicating risk to the public. Environmental Science and Technology 26:2048-2055.

Morgan, M. G., and M. Henrion. 1990. Uncertainty: A Guide to Dealing with Uncertainty in Quantitative Risk and Policy Analysis. New York, N.Y.: Cambridge University Press.

Moser, D. A. 1997. The Use of Risk Analysis by the U. S. Army Corps of Engineers. In Proceedings of a Hydrology and Hydraulics Workshop on Risk-Based Analysis for Flood Damage Reduction Studies. Davis, Calif.: U.S. Army Corps of Engineers Hydrologic Engineering Center.

Moser, D. A. 1998. Risk Analysis for Water Resources Planning and Management, PROSPECT Training Course Notes. Control No. 349. Washington, D.C.: U.S. Army Corps of Engineers.

NERC (Natural Environment Research Council). 1975. Hydrological Studies. Vol. 1 of Flood Studies Report. London: Natural Environmental Research Council.

NRC (National Research Council). 1983. Risk Assessment in the Federal Government: Managing the Process. Washington, D.C.: National Academy Press.

NRC. 1994. Science and Judgement in Risk Assessment. Washington, D.C.: National Academy Press.

NRC. 1995. Flood Risk Management and the American River Basin: An Evaluation. Washington, D.C.: National Academy Press.

NRC. 1996. Understanding Risk: Informing Decision in a Democratic Society. P. S. Stern, and Fineberg, H. V. (eds.). Washington, D.C.: National Academy Press.

NRC. 1999a. New Directions in Water Resources Planning for the U.S. Army Corps of Engineers. Washington, D.C.: National Academy Press.

NRC. 1999b. Downstream: Adaptive Management of Glen Canyon Dam and the Colorado River Ecosystem. Washington, D.C.: National Academy Press.

NWF (National Wildlife Federation). 1998. Higher Ground: A Report on Voluntary Property Buyouts in the Nation's Floodplains. Vienna, Vir.: National Wildlife Federation.

O'Leary, N. 1997. Risk Based Analysis of Beargrass Creek, Kentucky. Pp. 101-116 in Proceedings of a Hydrology and Hydraulics Workshop on Risk-Based Analysis for Flood Damage Reduction Studies. Davis, Calif.: U.S. Army Corps of Engineers Hydrologic Engineering Center.

Paté-Cornell, M. E. 1984. Warning systems: applications to the reduction of risk costs for new dams. Pp. 73-83 in J. L. Sarafim (ed.) Safety of Dams. Amsterdam: Balkema.

Pielke, R. 1999. Nine fallacies of floods. Climatic Change 42: 413-438.

Richards, F. 1999. A twentieth century history of flooding in the U.S. and what the future might hold. Pp. 144-151 in Second Symposium of Environmental Applications. Boston, Mass. American Meteorological Society.

Salas, J. D. 1993. Analysis and modeling of hydrologic time series. Chapter 19 in Maidment, D. R. (ed.). Handbook of Hydrology. New York, N.Y.: McGraw-Hill.

Stedinger, J. R. 1983a. Confidence intervals for design events. American Society of Civil Engineers Journal of Hydraulic Engineering 109 (HY1):13-27.

Stedinger, J. R. 1983b. Design events with specified flood risk. Water Resources Research 19(2):511-522.

Stedinger, J. R. 1997. Expected probability and annual damage estimators. Journal of Water Resources Planning and Management 123(2):125-35.

Stedinger, J. R., R. M. Vogel, and E. Foufoula-Georgiou. 1993. Frequency analysis of extreme events. Chapter 18 in Maidment, D. R. (ed.) Handbook of Hydrology. New York, N.Y.: McGraw-Hill, Inc.

Tasker, G. D., and J. R. Stedinger. 1986. Estimating generalized skew with weighted least squares regression. American Society of Civil Engineers Journal of Water Resources Planning and Management 112(2):225-237.

Tasker, G. D., and J. R. Stedinger. 1989. An operational GLS model for hydrologic regression. Journal of Hydrology 111(1-4): 361-375.

USACE. 1991a. Proceedings of a Hydrology and Hydraulics Workshop on Riverine Levee Freeboard. August 27-29, Monticello, Minnesota. Davis, Calif.: U. S. Army Corps of Engineers Hydrologic Engineering Center.

USACE. 1991b. Benefit Determination Involving Existing Levees. Policy Guidance Letter No. 26. Washington, D.C.: U.S. Army Corps of Engineers.

USACE. 1992a. Guidelines for Risk and Uncertainty Analysis in Water Resources Planning. Volume I: Principles with Technical Appendices. Report 92-R-1. Alexandria, Vir.: U.S. Army Corps of Engineers Institute for Water Resources.

USACE. 1992b. Guidelines and Procedures for Risk and Uncertainty Analysis in Corps Civil Works Planning. Volume II: Example

Cases, Guidelines for Risk and Uncertainty Analysis in Water Resources Planning. Report 92-R-2. Alexandria, Vir.: U.S. Army Corps of Engineers Institute for Water Resources.

USACE. 1996a. Risk-Based Analysis for Evaluation of Hydrology/Hydraulics, Geotechnical Stability, and Economics in Flood Damage Reduction Studies. ER 1105-2-101. Washington, D.C.: U.S. Army Corps of Engineers.

USACE. 1996b. Risk-Based Analysis for Flood Damage Reduction Studies. Manual, EM 1110-2-1619. Washington, D.C.: U.S. Army Corps of Engineers.

USACE. 1996c. An Introduction to Risk and Uncertainty Evaluation of Investments. Report 96-R-8. Alexandria, Vir.: U.S. Army Corps of Engineers Institute for Water Resources.

USACE. 1996d. Incorporating Risk and Uncertainty into Environmental Evaluation: An Annotated Bibliography. Report 96-R-9. Alexandria, Vir.: U.S. Army Corps of Engineers Institute for Water Resources.

USACE. 1997a. Risk and Uncertainty Analysis Procedures for the Evaluation of Environmental Outputs. Institute for Water Resources Report 97-R-7. Alexandria, Vir.: U.S. Army Corps of Engineers Institute for Water Resources.

USACE. 1997b. Hydrology and Hydraulics Workshop on Risk-Based Analysis for Flood Damage Reduction Studies. Publication SP-28. Davis, Calif.: U.S. Army Corps of Engineers Hydrologic Engineering Center.

USACE. 1997c. Metropolitan region of Louisville, Kentucky, Beargrass Creek Basin, Final Feasibility Report, Vol. 1, Main Report. Louisville, Kent.: U.S. Army Corps of Engineers.

USACE. 1997d. Metropolitan region of Louisville, Kentucky, Beargrass Creek Basin. Final Feasibility Report. Volume II: Technical Appendices. Louisville, Kent.: U.S. Army Corps of Engineers.

USACE. 1997e. Guidance on Levee Certification for the National Flood Insurance Program. CECW-P/CECW-E. 25 March 1997. Washington, D.C.: U.S. Army Corps of Engineers.

USACE. 1998a. Flood Reduction Studies for East Grand Forks, Minnesota and Grand Forks, North Dakota. Sub-Appendix A1, Hydrologic Analysis. St. Paul, Minn.: U.S. Army Corps of Engineers.

USACE. 1998b. Flood Reduction Studies for East Grand Forks, Minnesota and Grand Forks, North Dakota. Sub-Appendix A2, Hydraulic Analysis. St. Paul, Minn.: U.S. Army Corps of Engineers.

USACE. 1998c. Flood Reduction Studies for East Grand Forks, Minnesota and Grand Forks, North Dakota. Sub-Appendix A3, Risk – Based Analysis. St. Paul, Minn.: U.S. Army Corps of Engineers.

USACE. 1998d. HEC-FDA Flood Damage Reduction Analysis. Users Manual CPD-72, Version 1. Davis, Calif.: U.S. Army Corps of Engineers. Hydrologic Engineering Center.

USACE. 1999a. Digest of Water Resources Policies and Authorities. Washington, D.C.: U. S. Army Corps of Engineers.

USACE. 1999b. Tools for Risk Based Economic Analysis. Report 99-R-2. Alexandria, Vir.: U.S. Army Corps of Engineers Institute for Water Resources.

USACE. 1999c. Risk Analysis in Geotechnical Engineering for Support of Planning Studies. ETL 1110-2-556. Washington, D.C.: U.S. Army Corps of Engineers.

USACE. 2000. Planning Guidance Notebook. ER 1105-2-100. Washington, D.C.: U.S. Army Corps of Engineers.

U.S. Department of Agriculture (undated). Managing Risk: Being Prepared. Washington, D.C.: U.S. Department of Agriculture.

USWRC (U.S. Water Resources Council). 1973. Water and Related Land Resources: Establishment of Principles and Standards for Planning. Federal Register 38:24784, 248222-248223.

USWRC. 1983. Economic and Environmental Principles and Guidelines for Water and Related Land Resources Implementation Studies. Washington, D.C.: Water Resources Council.

Vanmarcke, E. 1983. Random Fields. Cambridge, Mass.: MIT Press.

Von Winterfeldt, D. and W. Edwards. 1986. Decision Analysis and Behavioral Research. New York, N.Y.: Cambridge University Press.

Wescoat, J. L. 1986. Expanding the range of choice in water resources management: An evaluation of policy approaches. Natural Resources Forum 10(3):239-254.

White, G. F. 2000. Water science and technology: Some lessons from the 20th century. Environment 42(1):30-38.

Yen, B. C., and Y. K. Tung. 1993. Some recent progress in reliability analysis for hydraulic design. Pp. 35-79 in Yen, B. C., and Y. K. Tung (eds.) Reliability and Uncertainty Analyses in Hydraulic Design. Reston, Vir.: American Society of Civil Engineers.

Yoe, C. E., and K. D. Orth. 1996. Planning Manual. Report 96-R-21. Alexandria, Vir.: U.S. Army Corps of Engineers Institute for Water Resources.

Appendix A

Glossary

Aggregated stage–damage function with uncertainty – A composite median stage–damage function for a damage reach. The function is developed by damage categories at the damage reach index location. The stage–damage functions of individual structures are aggregated using a series of water surface profiles to account for the slope in the profiles throughout the reach. Uncertainty, the error in the damage estimates, may also be computed.

Aleatory uncertainty – See "natural variability."

Annual exceedance probability – The probability that flooding will occur in any given year considering the full range of possible annual flood discharges.

Base flood – The median flood discharge having a 1 percent chance of being equaled or exceeded in any given year.

Bulletin 17B – A U.S. Geological Survey publication entitled *Guidelines for Determining Flood Flow Frequency* (USGS, 1982). The publication describes procedures for developing discharge–frequency functions using stream flow records. These procedures are recommended for all federal agency applications.

Conditional nonexceedance probability – The probability that failure will not occur during a flood of a given frequency. For example, a levee may have a 90 percent chance of not being overtopped when exposed to a 100-year flood.

Confidence limit curves – *Error limit curves* about a log-Pearson Type III discharge–probability function developed using the noncentral *t* distribution. Confidence limit curves are used to define the discharge–exceedance probability function's uncertainty.

Design flood – The flood that a flood damage reduction project, such as a levee, is based upon. Often the 100-year flood.

Discharge–exceedance probability – The relationship of peak discharge to the probability of that discharge being exceeded in any given year.

Egress – The ability to evacuate an area threatened by flood.

Epistemic uncertainty – See "knowledge uncertainty."

Equivalent record length – Number of years of a systematic record of recorded peak discharges at a stream gage. For probability functions derived at ungaged locations using model or other data, the equivalent record length is estimated based on the overall "worth" or "quality" of the frequency function expressed as the number of years of record. This parameter is important in risk-based analysis because it relates directly to the uncertainty of the flood–discharge probability function.

Exceedance probability event – The probability that a specific event will be equalled or exceeded in any given year. For example, the 0.01 exceedance probability event has 1 chance in 100 of occurring in any given year.

Expected annual damage – In risk-based analysis, the average or mean of all possible values of damage determined by Monte Carlo sampling of discharge–exceedance probability, stage–discharge, and stage–damage relationships and their associated uncertainties. Calculated as the integral of the damage–probability function.

FEMA – U.S. Federal Emergency Management Agency. FEMA administers the National Flood Insurance Program (NFIP) and is jointly responsible (with the U.S. Army Corps of Engineers) for levee certification within the NFIP.

FIA – Flood Insurance Administration. The federal entity within the Federal Emergency Management Agency responsible for administering the National Flood Insurance Program.

Flood damage reduction actions – Measures and actions taken to reduce flood damage. These may include implementation of reservoirs, detention storage, channels, diversions, levees and floodwalls, interior systems, flood-proofing, raising, relocation, and flood warning and preparedness actions.

Flood–frequency curve – A graph showing the average interval time (or recurrence interval) within which a flood of a given magnitude will be equaled or exceeded in any given year.

Freeboard – An addition to a levee's design height to ensure against overtopping during the design flood.

Hydrology and hydraulics – Hydrology involves the estimation of the amount and shape of the runoff–discharge hydrographs throughout the study area. It also includes the frequency of the events. Hydraulics involves analysis of stream water surface profiles, flood inundation boundaries, and other technical studies of stream flow characteristics.

Knowledge uncertainty – Uncertainty arising from imprecision in analysis methods and data.

Level of protection – A measure in years of the average interval between failures of a flood prevention system such as a levee.

Log normal distribution – A two-parameter probability distribution defined by the mean and standard deviation. A nonsymmetrical distribution applicable to many kinds of data sets where the majority (more than half) of values are less than the mean but where values greater than the mean can be extreme, such as with stream flow data.

Monte Carlo analysis – A method that produces a statistical estimate of a quantity by taking many random samples from an assumed probability distribution, such as a normal distribution. The method is typically used when experimentation is infeasible or when the actual input values are difficult or impossible to obtain.

Natural variability – Uncertainty arising from variations inherent in the behavior of natural phenomena (e.g., severity of the maximum flood in any year).

NED – National Economic Development. The water resources project planning alternative designed to maximize national economic development, consistent with protecting the nation's environment, and pursuant to national environmental statutes, applicable executive orders, and other federal planning requirements. The NED alternative is required by the *Principles and Guidelines* (*P&G*, see below) to be identified in Corps feasibility studies.

NFIP – National Flood Insurance Program, enacted by the federal government in 1968 to provide flood insurance for communities and structures at risk of flooding.

Nonstructural measures – Measures such as raising, relocating, floodproofing, and regulatory and emergency actions associated with structures and damageable property that modify the existing and/or future damage susceptibility. Nonstructural measures are not designed to directly affect the flow of flood waters.

Normal distribution – A two-parameter probability distribution defined by the mean and standard deviation. A symmetrical "bell shaped" curve applicable to many kinds of data sets where values are equally as likely to be greater than and less than the mean. Also called the Gaussian distribution.

One-hundred-year flood – A median flood discharge having a 1 percent chance of being equaled or exceeded in any given year.

P&G – Principles and Guidelines. A 1983 U.S. Water Resources Council document that provides water resource project planning guidance to the U. S. Corps of Engineers, the Bureau of Reclamation, the Natural Resources Conservation Service, and the Tennessee Valley Authority.

Probability function – A discharge–exceedance or stage–exceedance probability relationship for a reach developed by traditional, site-specific, hydrologic engineering analysis procedures.

Residual risk – The portion of the flood risk that still exists with the flood damage reduction project implemented.

Return period – The average time interval between occurrences of a hydrological event of a given or greater magnitude, usually expressed in years.

Risk – The probability of failure during a flood event. For reaches without levees, failure means exceeding a target stage. For reaches with levees, it means a levee failure.

Skewness coefficient – A statistic used as a measure of the symmetry of the statistical distribution of data. It is the third moment of a distribution. It is estimated as the number of values times the sum of the cubes of the deviations from the mean divided by the number of values minus 1, times the number of values times 2, times the standard deviation cubed.

Stage – The vertical distance in feet (meters) above or below a local or national elevation datum.

Stage associated with the median 1-percent chance flood discharge – The stage taken from the stage–discharge curve that corresponds to a discharge taken from the discharge–probability curve of 1 percent.

Stage–damage function – Relationship of depth of water to damage at a structure. Damage is normally specified as a percentage of the structure or content value. The functions are generic for similar structures and are not tied to the structure location.

Stage–damage functions with uncertainty – Stage–damage functions with uncertainty are computed at each structure and aggregated by damage category to damage reach index locations. Stage is elevation or locally referenced stage associated with the structure and index location. Damage is the median estimate of structure, content, and other inundation reduction damage associated with the stage of floodwaters at the location. Uncertainty in the stage–damage function arises from to errors in estimating the depth–damage function, first-floor stage, structure value, and content-to-structure-value ratio.

Stage–discharge function – A graphical relationship that yields the stage for a given discharge at a specific location on a stream or river.

Referred to as a rating function or curve. These relationships are usually developed by computing water surface profiles for several discharges and plotting the stages vs. discharge relationship at a specific stream location.

Stage–discharge functions with uncertainty – Relationship of the water surface stage and discharge. Uncertainty is the distribution of the errors of stage estimates about a specific discharge.

Standard deviation – A statistical measure of the spread of a distribution around the mean.

Structural measures – Those water resources project measures designed to modify the flow of flood waters.

Uncertainty – A measure of the imprecision of knowledge of variables in a project plan.

Appendix B

Corps–FEMA Levee Certification Documentation

DEPARTMENT OF THE ARMY
U.S. Army Corps of Engineers
WASHINGTON, D.C. 20314-1000

REPLY TO
ATTENTION OF:

Policy and Planning Division   8   AUG 1992

Mr. John L. Matticks
Assistant Administrator
Office of Risk Assessment
Federal Emergency Management Agency
500 C Street, SW.
Washington, DC 20472

Dear Mr. Matticks:

    The Army Corps of Engineers has developed new guidance
titled: "Risk Analysis Framework For Evaluation of
Hydrology/Hydraulics and Economics In Flood Damage Reduction
Studies." Once adopted, the procedures outlined in that document
will be used in the Corps flood damage reduction studies. This
recently completed effort, in the form of an Engineering
Circular, has been transmitted to our field offices for their
final review and comment.

    As you recall, this effort began as a result of discussions
at the August 1991 Hydrology and Hydraulics Workshop in
Riverwood, Minnesota. Your participation in that workshop was
extremely beneficial and provided a much needed FEMA perspective.

    Because of the Corps extensive involvement in and support of
the National Flood Insurance Program, and the potential
significant impact of this new guidance on Corps methods for
hydrologic and hydraulics analyses, your review of this document
is important to us. I have enclosed copies for your use and
would appreciate any comments you may have.

    Mr. Jerry Peterson of my staff will contact you in the near
future regarding the need for a meeting to discuss the new
guidance in further detail. Feel free to contact Mr. Peterson if
you have any questions in the interim.

                          Sincerely,

                          Hugh E. Wright

                          Jimmy F. Bates
                          Chief, Policy and Planning Division
                          Directorate of Civil Works

Enclosures

Federal Emergency Management Agency
Washington, D.C. 20472

APR 23 1993

CERTIFIED MAIL
RETURN RECEIPT REQUESTED

Mr. Jimmy F. Bates
Chief, Policy and Planning Division
Directorate of Civil Works
Department of the Army
U.S. Army Corps of Engineers
Washington, DC 20314-1000

This letter was simply forwarded to Corps field offices for guidance in performing levee certification in RBA studies.
DWD 5/18/99

Dear Mr. Bates:

This is in response to your letter dated August 3, 1992, regarding new guidance and procedures outlined in your draft circular entitled, "Risk Analysis Framework for Evaluation of Hydrology / Hydraulics and Economics in Flood Damage Reduction Studies," dated June 30, 1992. Thank you for allowing us to review and provide comments on this draft circular. We particularly appreciate the very informative presentation conducted by the U.S. Army Corps of Engineers (ACE) at our offices on March 2, 1993. By having the opportunity to ask questions, Federal Insurance Administration personnel gained a better understanding of the new procedures. Please relay our thanks to Messrs. Jerome Peterson, Earl Eiker, Bob Daniel, and Ken Zwickl of your staff.

We recognize that the new procedures contained in this circular are designed to account for the uncertainties inherent in hydrologic and hydraulic analyses when determining risk of failure of a flood control structure. As you know, our particular interest is whether we should recognize, on our maps, that a flood control structure provides protection from the base flood. Our regulations define the base flood as the condition of inundation having a one-percent chance of being equalled or exceeded during any given year. We have enclosed copies of the pertinent portions of our regulations for your convenience.

Our policy has been to recognize on our maps that a levee provides protection from the base flood if the ACE certifies that the levee provides that level of protection. It has been suggested that implementing the new procedures would restrict future certifications by the ACE by qualifying the level of protection in terms of reliability. For example, the ACE could determine that a levee will safely convey the 100-year discharge with 90-percent reliability. Contrary to

2

that suggestion, we believe the circular allows the ACE to be quite explicit. The new procedures determine the value that fills the blank in the following statement:

The levee has a _____ percent chance of being overtopped in any given year.

If the value is equal to or greater than 1.0 then we do not credit the levee on our maps; if it is less than 1.0 then we do credit the levee on our maps.

Traditionally, the term base flood has been synonymous with 100-year discharge. Because the methods used to define the base flood elevations essentially are one-to-one relationships between the discharge and elevation, no distinction was necessary between the base flood and the 100-year discharge. However, accounting for the uncertainties inherent in hydrologic and hydraulic analyses as described in the draft circular eliminates that convenient relationship. That is, a distinction must be made between the base flood and the 100-year discharge.

Because our definition of a base flood does not depend on a particular discharge, we believe that the ACE can certify that a levee has been adequately designed and constructed to provide protection from the base flood without the aforementioned qualification. The basis of such a certification would be a determination that the levee is structurally sound and everywhere higher than the elevations determined to have a one-percent chance of being equalled or exceeded in any given year. That would be the elevation corresponding to a simulation exceedance (true) probability of 0.01 in Table 3 of Appendix B to the draft circular.

In addition to the Federal Emergency Management Agency (FEMA), other programs, particularly at the State and local levels, will be affected by your proposed procedures. At the annual meeting of the Association of State Floodplain Managers (ASFPM) in Atlanta, Georgia, in March 1993, many officers of the ASFPM expressed a concern about the ACE's new risk-based analyses for levee certification. Additionally, several other members of the ASFPM with responsibility for floodplain management activities at the State and local levels shared similar concerns with members of my staff. It was explained that many of the states have adopted legislation or regulations that require the State and local bodies of government to assure that a specific number of feet of freeboard exists before a levee can be accredited. These ASFPM officers and members were very concerned that the new ACE procedures would put them in conflict with their own State and local statutes. I would urge you to obtain public comments on the ACE procedures before you finalize this engineering circular. Perhaps a briefing of the ASFPM comparable to that which you conducted in our offices would be of great benefit. Or, you may want to consider proposed and final rule-making with a public comment period to adopt the circular as ACE regulations. Otherwise, I fear you may be a target for some undesired congressional intercession.

We trust that this letter clarifies how your new procedures fit into the National Flood Insurance Program and, in particular, that the ACE can still certify levees for flood insurance purposes

3

without qualifying that certification. Again, thank you for the opportunity to review and comment on this draft circular. Mr. William R. Locke of my staff will serve as liaison during the implementation phase of these new procedures outlined in the draft circular. If you have any questions, you may contact him either by telephone at (202) 646-2754 or by facsimile at (202) 646-3445.

Sincerely,

John L. Matticks
Assistant Administrator
Office of Risk Assessment
Federal Insurance Administration

Enclosure

cc: Mr. Jerome Peterson
        Chief, Floodplain Management Services
           and Coastal Resources Branch
        U.S. Army Corps of Engineers

Federal Emergency Management Agency

Washington, D.C. 20472

MAR 21 1996

Major General Stanley G. Genega
Director of Civil Works
Department of the Army
U.S. Army Corps of Engineers
Washington, DC   20314-1000

Dear General Genega:

This is regarding our letter to the U.S. Army Corps of Engineers (USACE) dated April 23, 1993, concerning our comments on your draft circular entitled *Risk Analysis Framework for Evaluation of Hydrology/Hydraulics and Economics in Flood Damage Reduction Studies*, dated June 30, 1992. We had the opportunity to meet with USACE staff on September 13, 1995, to discuss the Federal Emergency Management Agency's (FEMA's) re-evaluation on Risk-Based Analysis with respect to FEMA's current levee certification policy.

It has come to my attention that the USACE has adopted a new levee design policy predicated on a risk-based approach. This risk-based analysis no longer incorporates freeboard as a design parameter.

In May 1995, we sent you a letter stating we were reviewing our position on Risk-Based Analysis as it pertains to our current levee certification policy and the comments in our April 1993 letter. As part of this ongoing review, a member of our staff attended a week-long training course in Risk-Based Analysis procedure. In addition, we reviewed the final results of analyses for 12 USACE projects provided to us by your staff. Based on this initial review, it became apparent that the criteria discussed in our April 1993 letter may not, in themselves, be the most appropriate standard for use by the USACE when certifying to FEMA that a levee may be credited for the purposes of removing the Special Flood Hazard Area (SFHA) designation from areas protected by that levee.

Our April 1993 letter pointed out that when using the Risk-Based Analysis procedures, the USACE could certify to FEMA that a levee provides protection from a base (1-percent) or less frequent flood without the use of a percent reliability. That letter was not intended to indicate that the USACE must certify levees to

2

FEMA if the simulation exceedance (true) probability was 0.01 or less. For the 12 USACE projects, the simulation exceedance (true) probability standard of 0.01 referenced in our April 1993 letter produced levee designs with only 0.1 to 1.5 feet of freeboard and contained the FEMA base flood with a reliability of between only 50 and 75 percent. I am concerned that there may be a potential for conflict between levee projects designed by the USACE and by other Federal Agencies and Private Engineering firms. It is conceivable that levee designs (especially levee heights) would differ significantly between the USACE and others, although both would be designed to provide protection against the one-percent chance event. At this point, we are asking that the USACE continue to use the engineering and judgment expertise you are known for when certifying levees to FEMA, and not rely merely on the simulation exceedance probability if that, in the opinion of the USACE, results in unacceptable freeboard height.

Although we are rescinding the April 1993 letter and restoring the previous certification criteria, I believe that it would be in the best interests of FEMA and the National Flood Insurance Program to continue dialoguing on this issue. Until such time as detailed criteria can be developed (which may include the concepts of annual exceedance probability and reliability), FEMA will continue to accept letters of certification from the USACE stating that a particular levee has been adequately designed and constructed to provide protection against the FEMA base flood as a means of removing the SFHA designation from areas behind the levee.

I look forward to discussions with you on this matter and other issues of mutual interest.

Sincerely,

Richard W. Krimm
Acting Associate Director
Mitigation Directorate

cc: Jerome Peterson, USACE

DEPARTMENT OF THE ARMY
U.S. Army Corps of Engineers
WASHINGTON, D.C. 20314-1000

REPLY TO
ATTENTION OF:

05 DEC 1996

RECEIVED

07 APR 1997

HFC

Planning Division
Flood Plain Management Services
and Coastal Resources Branch

Mr. Richard W. Krimm
Associate Director
Mitigation Directorate
Federal Emergency Management Agency
Washington, DC 20472

Dear Mr. Krimm:

This concerns your letter of March 21, 1996, and our recent discussions regarding the use of risk-based analyses in flood damage reduction project planning and levee certification for National Flood Insurance Program (NFIP) purposes.

I strongly believe that the risk-based approach to project formulation provides important information on potential levee performance that should be utilized in levee certification decisions. As we have discussed, our desire to introduce these concepts into levee certification decisions must be tempered with your need to maintain consistency with existing NFIP regulations. With that in mind, we have developed specific guidance to ensure consistent application of engineering principles and judgement when the U.S. Army Corps of Engineers provides levee certification information to your agency in support of the NFIP.

I would appreciate your comments on the enclosed guidance. It is our intent to provide the guidance to our districts and divisions as soon as possible to eliminate any confusion that may exist on this issue.

Sincerely,

John P. D'Aniello, P.E.
Deputy Director of Civil Works

Enclosure

DEPARTMENT OF THE ARMY
U.S. Army Corps of Engineers
WASHINGTON, D.C. 20314-1000

. 1 0 APR 1997

REPLY TO
ATTENTION OF:

CECW-P/CECW-E

MEMORANDUM FOR ALL MAJOR SUBORDINATE COMMANDS

SUBJECT: Guidance on Levee Certification for the National Flood Insurance Program

1. Use of risk-based analysis by the U.S. Army Corps of Engineers in flood damage reduction project formulation studies has created a disconnect between the Corps analysis and the Federal Emergency Management Agency's (FEMA) levee certification policy. FEMA's policy requires that levees be structurally sound, properly maintained, and have at least three feet of freeboard above the 100-year flood profile elevations before FEMA will recognize that the levees provide protection. The Corps risk-based analysis eliminates the concept of arbitrary freeboard by incorporating risk and uncertainty throughout the formulation process.

2. To ensure that levee certification to FEMA is performed by the Corps in a consistent manner, the enclosed guidance has been developed for use by all Major Subordinate Commands (MSC). This guidance has been reviewed and accepted by FEMA, and establishes Corps-wide standard procedures applicable to all future levee certification decisions.

3. It is recognized that levee certification commitments based on existing FEMA regulations have been made to non-Federal sponsors for some projects in progress. Exceptions to the new guidance will be considered for uncertified projects for which levee certification commitments already have been made. Each MSC should submit a list of projects that fall into this category, along with a justification for the exception, to CECW-EH by NLT 30 April 1997.

4. Points of contact for this guidance are Mr. Earl Eiker, telephone (202) 761-8500, or Mr. Ken Zwickl, telephone (202) 761-1855.

FOR THE COMMANDER:

Encl

RUSSELL L. FUHRMAN
Major General USA
Director of Civil Works

DISTRIBUTION: See next page.

CECW-P/CECW-E   **25 March 1997**

GUIDANCE ON LEVEE CERTIFICATION
FOR THE
NATIONAL FLOOD INSURANCE PROGRAM

1. **PURPOSE AND APPLICABILITY:** This document provides guidance to be used for certifying levees to the Federal Emergency Management Agency (FEMA) for their administration of the National Flood Insurance Program (NFIP). This guidance does not affect plan formulation and evaluation procedures. It is intended to provide a consistent methodology for levee certification by the Corps of Engineers. This guidance applies to all Corps District and Division offices. Note that levee certifications are provided to FEMA at the District/Division option and within available funds.

2. **BACKGROUND:** By letter dated 21 March 1996, FEMA requested that the Corps review its criteria for levee certification in order to ensure consistency in administration of the NFIP by FEMA. This concern has arisen as a result of the Corps application of Risk-Based Analysis (RBA) in flood damage reduction project formulation studies. FEMA's policy requires that levees be structurally sound, properly maintained, and have at least 3 feet of freeboard above the 100-year flood profile elevations before FEMA will recognize that the levees provide protection from the 100-year flood. The FEMA requirements are fully explained in 44 CFR, Chapter 1, Part 65.10 of the Code of Federal Regulations. The FEMA requirements include data and analysis submission requirements for design criteria (freeboard, closures, embankment protection, embankment and foundation stability, settlement, interior drainage), operations plans and maintenance plans. 44 CFR Part 65.10 also states that in lieu of the structural requirements and data and analysis requirements, a Federal agency with responsibility for levee design may certify that a levee has been adequately designed and constructed to provide 100-year protection.

Levee certification for NFIP purposes can best be explained as follows. FEMA may request a "levee certification" from the Corps by letter directly to the Corps District office. The letter normally contains language such as:

> "...Please provide this office with current certification as to whether the design and maintenance of this levee are adequate to credit it with 100-year flood protection. Please note that such a statement does not constitute a warranty of performance, but rather the Corps current position of the levee system's design adequacy..."

3. **POLICY:** The Corps will continue to work with FEMA to ensure that Risk-Based Analysis provides improved information for levee certification decisions. The following guidance and decision tree should be used until further notice.

CECW-P/CECW-E   25 March 1997
GUIDANCE ON LEVEE CERTIFICATION
FOR THE NATIONAL FLOOD INSURANCE PROGRAM

a. **Existing Levees, No Risk-Based Analysis Available:** For certification purposes, the Corps should evaluate the levees based primarily on FEMA criteria contained in 44 CFR Chapter 1, Part 65.10. Thus, the general rule will be that if a levee will contain the median one percent chance flood, with three feet of freeboard, it should be certified as being capable of passing the FEMA base flood, as long as it is adequate based on a geotechnical and structural evaluation, as described below. Exceptions to the three feet of freeboard requirement may be pursued, based on the FEMA policy of permitting other Federal agencies responsible for levee construction to certify that levees will pass the FEMA base flood. Such exceptions should be based on careful evaluation of the hydrologic, hydraulic, structural and geotechnical uncertainties, and current levee condition as discussed below.

b. **Existing and Proposed Levees, Risk Based Analysis Available:** In these cases, output on project performance from the Risk-Based Analysis should be used to arrive at a decision regarding levee certification for FEMA. Existing and proposed levees will be certified as capable of passing the FEMA base flood if the levees meet the FEMA criteria of 100-year flood elevation plus three feet of freeboard, with two exceptions, as follows. When the FEMA criteria results in a "Conditional Percent Chance Non-exceedance" (Reliability) of less than 90%, the minimum levee elevation for certification will be that elevation corresponding to a 90% chance of non-exceedance. When the FEMA criteria results in a reliability of greater than 95%, the levee may be certified at the elevation corresponding to a 95% chance of non-exceedance. For existing levees, the certification decision is also contingent upon a structural and geotechnical evaluation, as described below. For proposed levees, the geotechnical and structural issues are assumed to be accounted for during design and construction of the levees.

c. **Engineering Evaluation:** A geotechnical and structural evaluation will be used to determine the water elevation at which the levee is not likely to fail. In some cases, this water level will be the determining factor in the decision to certify the levee system. The procedures to be used in the evaluation of a levee system for NFIP levee certification should consist of an engineering evaluation to determine if the levee system meets the Corps design, construction, operation and maintenance standards, regardless of levee ownership or responsibility. The District will examine available existing information and data, such as original design, surveys of levee top profile, levee cross-sections, records of modifications and changes, performance during past flood events, and remedial measures. It will also include a field inspection of the levee, structures, closure devices and pumping stations to evaluate the adequacy of maintenance. The engineering analysis should examine the project with respect to embankment stability, underseepage, through seepage, and erosion protection. Existence of closure devices will necessitate a review of the adequacy of flood warning time for the complete operation of all closure structures.

(Page 2 of 3)

LEVEE CERTIFICATION DECISION TREE

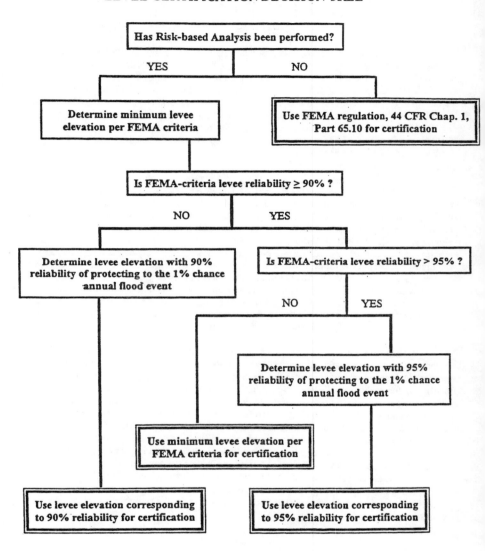

FEMA Criteria = 1% chance median annual flood event plus three feet of freeboard
RELIABILITY = % chance non-exceedance given the 1% chance annual event occurs

(Page 3 of 3)

Appendix C

Economic and Environmental Principles for Water Related Land Resources Implementation Studies

Economic and Environmental Principles for Water and Related Land Resources Implementation Studies

These Principles are established pursuant to the Water Resources Planning Act of 1965 (Pub. L. 89-80), as amended (42 U.S.C. 1962a-2 and d-1). These Principles supersede the Principles established in connection with promulgation of principles, standards and procedures at 18 CFR, Parts 711, 713, 714 and 716.

1. Purpose and Scope

These principles are intended to ensure proper and consistent planning by Federal agencies in the formulation and evaluation of water and related land resources implementation studies.

Implementation studies of the following agency activities are covered by these principles:

(a) Corps of Engineers (Civil Works) water resources project plans;

(b) Bureau of Reclamation water resources project plans;

(c) Tennessee Valley Authority water resources project plans;

(d) Soil Conservation Service water resources project plans.

Implementation studies are pre- or postauthorization project formulation or evaluation studies undertaken by Federal agencies.

2. Federal Objective

The Federal objective of water and related land resources project planning is to contribute to national economic development consistent with protecting the Nation's environment, pursuant to national environmental statutes, applicable executive orders, and other Federal planning requirements.

(a) Water and related land resources project plans shall be formulated to alleviate problems and take advantage of opportunities in ways that contribute to this objective.

(b) Contributions to national economic development (NED) are increases in the net value of the national output of goods and services, expressed in monetary units. Contributions to NED are the direct net benefits that accrue in the planning area and the rest of the Nation. Contributions to NED include increases in the net value of those goods and services that are marketed, and also of those that may not be marketed.

3. State and Local Concerns

Federal water resources planning is to be responsive to State and local concerns. Accordingly, State and local participation is to be encouraged in all aspects of water resources planning. Federal agencies are to contact Governors or designated State agencies for each affected State before initiating studies, and to provide appropriate opportunities for State participation. It is recognized, however, that water projects which are local, regional, statewide, or even interstate in scope do not necessarily require a major role for the Federal Government; non-Federal, voluntary arrangements between affected jurisdictions may often be adequate. States and localities are free to initiate planning and implementation of water projects.

4. International Concerns

Federal water resources planning is to take into account international implications, including treaty obligations. Timely consultations with the relevant foreign government should be undertaken when a Federal water project is likely to have a significant impact on any land or water resources within its territorial boundaries.

5. Alternative Plans

Various alternative plans are to be formulated in a systematic manner to ensure that all reasonable alternatives are evaluated.

(a) A plan that reasonably maximizes net national economic development benefits, consistent with the Federal objective, is to be formulated. This plan is to be identified as the NED plan.

(b) Other plans which reduce net NED benefits in order to further address other Federal, State, local, and international concerns not fully addressed by the NED plan should also be formulated.

(c) Plans may be formulated which require changes in existing statutes, administrative regulations, and established common law; such required changes are to be identified.

(d) Each alternative plan is to be formulated in consideration of four criteria: completeness, effectiveness, efficiency, and acceptability. Appropriate mitigation of adverse effects is to be an integral part of each alternative plan.

(e) Existing water and related land resources plans, such as State water resources plans, are to be considered as alternative plans if within the scope of the planning effort.

6. Plan Selection

A plan recommending Federal action is to be the alternative plan with the greatest net economic benefit consistent with protecting the Nation's environment (the NED plan), unless the Secretary of a department or head of an independent agency grants an exception to this rule. Exceptions may be made when there are overriding reasons for recommending another plan, based on other Federal, State, local and international concerns.

7. Accounts

Four accounts are established to facilitate evaluation and display of effects of alternative plans. The national economic development account is required. Other information that is required by law or that will have a material bearing on the decision-making process should be included in the other accounts, or in some other appropriate format used to organize information on effects.

(a) The national economic development (NED) account displays changes in the economic value of the national output of goods and services.

(b) The environmental quality (EQ) account displays nonmonetary effects on significant natural and cultural resources.

(c) The regional economic development (RED) account registers changes in the distribution of regional economic activity that result from each alternative plan. Evaluations of regional effects are to be carried out using nationally consistent projections of income, employment, output, and population.

(d) The other social effects (OSE) account registers plan effects from perspectives that are relevant to the planning process, but are not reflected in the other three accounts.

8. Discount Rate

Discounting is to be used to convert future monetary values to present values.

9. Period of Analysis

The period of analysis to be be the same for each alternative plan.

10. Risk and Uncertainty

Planners shall identify areas of risk and uncertainty in their analysis and describe them clearly, so that decisions can be made with knowledge of the degree of reliability of the estimated benefits and costs and of the effectiveness of alternative plans.

11. Cost Allocation

For allocating total project financial costs among the purposes served by a plan, separable costs will be assigned to their respective purposes, and all joint costs will be allocated to purposes for which the plan was formulated. (Cost sharing policies for water projects will be addressed separately.)

12. Planning Guidelines

In order to ensure consistency of Federal agency planning necessary for purposes of budget and policy decisions and to aid States and the public in evaluation of project alternatives, the Water Resources Council (WRC), in cooperation with the Cabinet Council on Natural Resources and Environment, shall issue standards and procedures, in the form of guidelines, implementing these Principles. The head of each Federal agency subject to this order will be responsible for consistent application of the guidelines. An agency may propose agency guidelines which differ from the guidelines issued by WRC. Such agency guidelines and suggestions for improvements in the WRC guidelines are to be submitted to WRC for review and approval. The WRC will forward all agency proposed guidelines which represent changes in established policy to the Cabinet Council on Natural Resources and Environment for its consideration.

13. Effective Date

These Principles shall apply to implementation studies completed more than 120 days after issuance of the standards and procedures referenced in Section 12, and concommitant repeal of 18 CFR, Parts 711, 713, 714, and 716.

These economic and environmental Principles are hereby approved.

Ronald Reagan

February 3, 1983

Appendix D

Functions of Random Variables

A random variable X is a variable whose probability of taking on a particular value x in an infinitesimal range is described by a probability density function, f(x). The mean or expected value of X is given by

$$E(X) = \mu_x = \int_{-\infty}^{\infty} xf(x)dx, \qquad (1)$$

in which the product $f(x)dx$ is the probability of x occurring in an interval $[x, x + dx]$. The variance, σ_x^2, is similarly

$$\sigma_x^2 = E(x - \mu_x)^2 = \int_{-\infty}^{\infty} (x - \mu_x)^2 f(x)dx. \qquad (2)$$

When Monte Carlo simulation of a random variable is carried out, a set of n independent values is generated to yield a set of replicates $\{x_1, x_2, \ldots, x_n\}$, from which the mean is estimated as

$$\bar{x} = \sum_{i=1}^{n} \frac{1}{n} x_i = \frac{1}{n} \sum_{i=1}^{n} x_i. \qquad (3)$$

The weight, $1/n$, implies each value is as likely as any other. Equation 3 represents the process actually used in the Corps's risk analysis procedure, in that the weight, $1/n$, approximates the theoretical probability, $f(x)dx$, and the summation in Equation 3 replaces the integral in

Equation 1.

When a sum Z of two random variables, X and Y, is required, the process is more complex. For two variables, x and y, the corresponding z is

$$z = x + y,\tag{4}$$

and the expected value of Z is the sum of the expected values of X and Y:

$$\mu_z = \mu_x + \mu_y.\tag{5}$$

However, the variance of Z is

$$\sigma_z^2 = \sigma_x^2 + \sigma_y^2 + 2\rho_{xy}\sigma_x\sigma_y,\tag{6}$$

where ρ_{xy} is the correlation coefficient of x and y $(-1 \le \rho_{xy} \le 1)$. The correlation coefficient introduces a new element into the picture and represents the degree of association of values of x and y. When the variables are statistically independent, $\rho_{xy} = 0$, and the variance of the sum is simply the sum of the variances. When the variables are positively correlated, the variance of the sum is increased by an amount proportional to the degree of correlation.

Similarly, when the difference, Z, between two random variables, X and Y, is found, the value of the variate z can be found as:

$$z = x - y\tag{7}$$

and the expected value as

$$\mu_z = \mu_x - \mu_y,\tag{8}$$

while the variance of the difference is given by

$$\sigma_z^2 = \sigma_x^2 + \sigma_y^2 - 2\rho_{xy}\sigma_x\sigma_y.\tag{9}$$

In this case, if the variables are positively correlated, the variance of the difference is diminished by an amount proportional to the degree of correlation.

The significance of all these definitions is that Monte Carlo simulation works at the level of replicates, or individually generated values of variables x and y. At that level, the normal rules of arithmetic for sums and differences apply, as specified by equations 4 and 7, and they can also be applied to the expected means of those variables, as given by Equations 5 and 8. However, the variability of a sum or difference of random variables depends in part on the variability in the individual variables and also on the degree of correlation or interdependence between the variables. Properly quantifying variability in a problem involving the interaction of several random variables requires an understanding and a correct representation of their interdependence or correlation.

Appendix E

Biographical Information

GREGORY B. BAECHER (*Chair*) is a professor in and the chair of the civil engineering program at the University of Maryland. Prior to joining the faculty at Maryland in 1995, Dr. Baecher served on the faculty of civil engineering at the Massachusetts Institute of Technology from 1976 to 1988, and he served as the CEO and founder of ConSolve Incorporated, Lexington, Massachusetts, from 1988–1995. His fields of expertise include risk analysis, water resources engineering, and statistical methods. Dr. Baecher received his B.S. degree in civil engineering from the University of California-Berkeley and his M.S. and his Ph.D. degrees in civil engineering from the Massachusetts Institute of Technology.

EFI FOUFOULA-GEORGIOU is a professor of civil engineering and the director of the St. Anthony Falls Laboratory, University of Minnesota. Her research focuses on understanding and modeling the complex spatio-temporal organization and interactions of hydrologic processes, including precipitation and landforms. Dr. Foufoula-Georgiou obtained her diploma in civil engineering from the National Technical University of Athens, Greece, and her Ph.D. degree in environmental engineering from the University of Florida. She has chaired and served on many national and international committees and government advisory panels and has served on the editorial boards of several journals.

RALPH KEENEY is a professor of business and a professor of systems engineering with the Center for Telecommunications Management at the University of Southern California. His areas of expertise include decision analysis, risk analysis, and management decision making. His experience includes large-scale siting studies, risk analysis, energy policy, and environmental studies. Groups to which he has served as a consultant include Seagate Technology, American Express, British Columbia Hydro, Pacific Gas and Electric, and the U.S. EPA. Dr. Keeney received his B.S. degree in engineering from the University of California, Los Angeles, his S.M. degree and his E.E. degree in electrical engineering from the Massachusetts Institute of Technology, and his Ph.D. degree in operations research from the Massachusetts Institute of Technology. Dr. Keeney is a member of the National Academy of Engineering.

LESTER LAVE is a professor of engineering and public policy in the engineering school and the Higgins Professor of Economics in the business school at Carnegie Mellon University. Dr. Lave's research has focused on the use of risk analysis as it relates to a range of health, safety, and environmental issues, including carcinogenic chemicals, natural resource valuation, and global climate change. He has served as a consultant to several federal and state government agencies, including the U.S. Environmental Protection Agency, Office of Safety and Health Administration, Human Health Services, as well as to many corporations, including General Motors and Xerox. Dr. Lave is a past member of the Water Science and Technology Board and is a member of the Institute of Medicine.

HARRY F. LINS is a hydrologist with the U.S. Geological Survey in Reston, Virginia. Dr. Lins's principal research interests are in the areas of hydroclimatology, surface water hydrology, and multivariate statistics. Dr. Lins has served since 1989 as the coordinator of the Global Change Hydrology Program of the Geological Survey's Water Resources Division. Dr. Lins received his B.S. degree from the University of Maryland, his M.S. degree from the University of Delaware, and his Ph.D. degree from the University of Virginia.

DANIEL P. LOUCKS is a professor of civil and environmental engineering at Cornell University. Dr. Loucks's primary research interests are in water resource systems planning and analysis, decision support systems, and applications of engineering methods to water and environmental problems. He has taught at a number of universities in the United States and abroad and has served as a consultant to both public and private sector organizations. He received his B.S. degree from Pennsylvania

State University, his M.F. degree from Yale University, and his Ph.D. degree from Cornell University. Dr. Loucks is a member of the National Academy of Engineering.

DAVID R. MAIDMENT is the Ashley H. Priddy Centennial Professor of Engineering and the Director of the Center for Research in Water Resources at the University of Texas, Austin. Dr. Maidment's primary areas of research and teaching include water resources engineering, geographic information systems, and statistical methods in hydrology and water resources. He received his bachelor of agricultural engineering degree (first-class honors) at the University of Canterbury, New Zealand, and his M.S. and Ph.D. degree in civil engineering from the University of Illinois.

MARTIN W. MCCANN is the president of Jack R. Benjamin and Associates, Inc., in Menlo Park, California. He is also a consulting professor of civil and environmental engineering at Stanford University, where he is the chair of the National Performance of Dams Program (NPDP). The NPDP, founded at Stanford by Dr. McCann, is a program that has created a national network to report dam incidents and to archive this information for use by the profession. Dr. McCann's professional background includes probabilistic hazards analysis, including hydrologic events, risk assessment, reliability analysis, uncertainty analysis, and systems analysis. Dr. McCann has been a consultant to several government and private sector groups in the United States and abroad. He received his B.S. degree from Villanova University and his M.S. and his Ph.D. degrees from Stanford University.

JERY R. STEDINGER is a professor of civil and environmental engineering at Cornell University. His research focuses on the optimal operation of reservoir systems, efficient use of hydrologic data, risk analysis, and stochastic hydrology. Dr. Stedinger has served on several NRC committees, including the Committee on Flood Control Alternatives in the American River Basin and the Committee on American River Flood Frequencies. He earned his B.A. degree in applied mathematics from the University of California, Berkeley and his M.S. and his Ph.D. degrees from Harvard University.

BEN CHIE YEN is a professor of water resources engineering at the University of Illinois at Urbana-Champaign since 1966. His areas of expertise include urban storm drainage, watershed hydrology, risk and reliability analysis, and open channel and river hydraulics. In the past 35 years he has been a consultant to public and private sectors in the United States and abroad, and he has been visiting or guest professor at a dozen universities in four continents. He received his B.S.C.E. degree from the National Taiwan University and his M.S. and Ph.D. degrees from the University of Iowa.

STAFF

JEFFREY W. JACOBS is a senior staff officer with the National Research Council's Water Science and Technology Board and served as this committee's study director. Dr. Jacobs's research interests include institutional and policy arrangements for water resources management and international cooperation in water development. He has studied these issues extensively in the Mekong River basin of Southeast Asia and has also conducted comparative studies in water policy in the Mekong and Mississippi River systems. Dr. Jacobs received his Ph.D. degree in geography from the University of Colorado-Boulder.

ELLEN A. DE GUZMAN is a senior project assistant at the National Research Council's Water Science and Technology Board. She received her B.A. degree from the University of the Philippines and is majoring in economics at the University of Maryland University College. She has worked with a number of studies including *Watershed Management for Potable Water Supply, Issues in Potable Reuse, Valuing Ground Water, New Directions in Water Resources Planning for the U.S. Army Corps of Engineers,* and *Improving American River Flood Frequency Analyses.*